D0209790

Nineteenth-Century
American Science

Nineteenth-Century American Science

A Reappraisal

GEORGE H. DANIELS, EDITOR

NORTHWESTERN UNIVERSITY PRESS

EVANSTON

1972

Framingham State College
Framingham, Massachusetts

Contents

Contents, *continued*

Introduction

Until quite recently, it has been customary for historians to assume that American science is a product of the twentieth century. Historians have considered America's first Nobel laureate in 1907, the mobilization for World War I, or even, in some accounts, the immigration of distinguished refugees from Nazi Germany in the 1930s as the beginning of American science. Pretwentieth-century developments have been summarized with passing mention of Benjamin Franklin, Joseph Henry, and Willard Gibbs. Only a decade ago, a student who wished to study science in America in the nineteenth century (at that time one had to use the circumlocution—"American science" was not permitted) was advised that there was nothing to study and that he should turn to science in France, Germany, or England for a subject.

Historians' distorted views of nineteenth-century American science are in large measure the result of their acceptance of the cries of neglect that came from the scientific community. As Howard Miller points out in his essay, nineteenth-century scientists constantly bemoaned America's inferior position in the scientific world in the hope that they could convince potential patrons that the cause of science was the cause of social progress and national prestige, and that it was therefore worthy of their support. Unfortunately, the scientists' letters and articles, while convincing an impressive number of public and private patrons, also convinced several twentieth-century scholars who have examined nineteenth-century science solely through the eyes of nineteenth-century scientists. There never seems to be enough money for science, or enough public appreciation of science, and nineteenth-century American scientists, like their British or twentieth-century American counterparts, tended to focus on the areas of neglect in their public laments. In private correspondence, however, one can find evidence that many scientists were aware of a growing appreciation of their work.

The tendency of scientists to disavow the past in the name of disciplinary distinctiveness, methodological refinement, and theoretical purity, which Robert Davis identifies in twentieth-century social scientists, is probably widely shared, and may have contributed to the distorted view. In this particular case, Davis found a rich tradition of statistical studies dating from the mid-1800s which, in his view, entirely invalidates the claim that the legacy of nineteenth-century social science was "merely one of speculative theory and utopian reforms." Davis' work suggests the profound differences in interpretation that may result when scholars look more carefully at the actual context of science.

Fortunately, the profession seems largely to have passed beyond this strange combination of self-deprecation, disciplinary self-serving, and the nationalism that is inherent in attempts to compare a sparsely populated, developing nation—as America was throughout most of the nineteenth century—with all of Europe. As Edward Lurie's essay indicates, books and articles on the history of American science have been appearing with increasing frequency and sessions have been devoted to it at recent meetings of the American Historical Association and the Organization of American Historians. The 1971 meeting of the American Association for the Advancement of Science included a session devoted to the social origins of American science. As a result of all of this work, a new picture of the formative period in American science is beginning to emerge.

The late nineteenth century, it is becoming evident, was a period of flux between the era of colonial dependency and the period of maturity. Specialist societies were being founded, publication outlets proliferated, a new leadership began to emerge, and unprecedented new opportunities arose for scientific work. The growth of colleges and universities throughout the country led to growing employment possibilities, as did the rise of scientific bureaus in the federal government. A revolution in transportation made it possible for scientists across the nation to organize effectively and to meet frequently enough to form a pressure group of growing power. What is more, the internal development of science had finally made it relevant to the industrial interests of the country, and the entrepreneurial climate provided opportunities for the application of science in a variety of areas. Emboldened by their new position in the society, scientists were beginning to drop their emphasis on practicality and to develop strong theoretical interests as well as a new ideal of "pure science" that justified their new interests. It was, in short, a period of growth and development during which the groundwork for American scientific dominance in the twentieth century was laid.

Each of the contributors to this volume has been instrumental in reaching the new understanding of American science. All are concerned in a major way with what is often (and perhaps unfortunately) termed "external" history—the study of science in relation to particular social settings. Unlike the "internalists," their focus is not on the scientific discipline, but on sets of scientific communities, which they see as groups of scientific workers sharing activities which have methodological, theoretical, and instrumental aspects, and interacting with other groupings in the society. A basic tenet is that science is a social activity, and most would share with Edward Lurie the conviction that the history of American science is interesting and important only if it is viewed as an especially significant factor in the history of American national development.

Beyond this general methodological agreement, the interests and approaches of the authors vary immensely, and in preparing papers for this volume they have been encouraged to follow their interests. The invitation, issued by Northwestern's Center for the Interdisciplinary Study of Science and Technology, specified only that each paper was to contribute to a "general reassessment of nineteenth-century American science." The papers quite naturally divide themselves into three categories, represented by the section titles in this volume. The first five papers supply a critique of the historiography of American science and offer a variety of general approaches to the subject. The second section offers case studies of scientists in their research or organizational roles. Papers in the third section deal with the relations among science, technology, and entrepreneurship, and suggest what may be the source of American uniqueness in science.

Each of the papers contributes in some measure toward the general "reassessment," and together they suggest the broad outlines that the new interpretation will have. Nathan Reingold points out in his provocative chapter that recent studies indicate that the research pattern in nineteenth-century America was not greatly different from the European norm, which has not yet been discerned. Recent historical studies invariably uncover more basic or pure research than was predictable from traditional accounts, and recognize more scientific work of an abstract, theoretical kind than was predictable from traditional accounts. These findings, Reingold believes, suggest that we must take a new and more critical look at the alleged "indifference to basic research." In Reingold's view, the concept of "indifference" has had a multitude of unfortunate consequences—an unwarranted emphasis on the colonial period, an unsuccessful search for the great men and great contributions, the contrived distinctions between pure and applied science, and an overemphasis on the physical sciences.

Mark Beach, exploring one instance of a common historical search for great men, concludes that the group known as the Lazzaroni actually made no real attempt to control the course of nineteenth-century American science. The group, headed by Henry, Bache, Agassiz, and Peirce, actually was of no consequence beyond the importance of the members as individuals. In Beach's well-argued view, the fact that historians have so long believed in the power of the "scientific Lazzaroni" is merely another instance of the hazards of the uncritical acceptance of contemporaneous statements, in this case statements of men who had been frustrated in one way or another in their professional aspirations during a period of turmoil and change.

Howard Miller, like Reingold, denies that Americans were necessarily uninterested in pure science and shows that there has been a long tradition in America of mixed public-private enterprise in support of science. With a few loud—albeit impotent—exceptions, American scientists have been happy to get their support from any source; and, as Miller shows, they were remarkably successful in the nineteenth century. Classical economic theory, Miller finds, was no inhibition to governmental funding of scientific expeditions and projects. When doctrinal arguments were raised against governmental support, as they were in the debates relating to the Smithsonian Institution, they were usually masks for other, more concrete objections. The major issues concerning the scientific community, he insists, such as the increase of technical knowledge, professionalization and institution-building, and the creation of a reliable source of support, were matters which paid no attention to a supposed conflict between public and private enterprise. In this case, as in the case of "American indifference," historians have too readily accepted the rhetoric of contemporaries as solid historical evidence.

Daniel Kevles, despite his caveat regarding *any* generalization about American attitudes toward scientific research, nevertheless finds a high level of enthusiasm among the intelligentsia and most social leaders. Physics was undersupported in the nineteenth century, but primarily because physicists had not yet found a common interest on which to erect a sound institutional base. In general, however, he reports a growing interest and appreciation among Americans during the last quarter of the nineteenth century.

A central difficulty, noted by Reingold, is that precise information regarding European attitudes toward science is lacking, as is knowledge of the typical patterns of research in Europe. Without this information, meaningful comparisons cannot be made, and students of European science have shown no inclination to provide it. With the exception of earlier Marxist writers, who were never very precise or convincing, and a group of sociologists including Robert K. Merton, Bernard Barber, and Joseph Ben-David, who have only partially filled in the blanks, students of European science have pursued an uncompromising "internalist" line, and their work is therefore of little value to the student of American science who is seeking to determine what is unique about science in American culture.

In the scientific communities that have been studied, applied work always appears to be a larger part of the total effort than basic research. As Reingold suggests, America has probably not been unique in this respect. Whatever the European situation, the distinction between "basic" and "applied" was so blurred in the context of nineteenth-century America that in most concrete research situations both were involved. Indeed, this may be characteristic of science at all times and in all places, but it was an especially marked feature of a society expanding rapidly into scientifically unexplored territory. Under these American conditions, "practically" oriented work often had clear "basic" implications. As Charles Rosenberg observes, in the public mind the functions of research and development were never clearly distinguished from those of regulation and inspection. While this confusion has caused a great deal of grief for some American scientists, it could nevertheless be exploited to great advantage by the alert research entrepreneur on behalf of his profession. Indeed, as Don K. Price pointed out in his early study of science and government, the American "unwillingness to take the answer from established authority leads to a tremendous use of research as a basis of decision in all areas."[1] Building upon this interest, scientists in a variety of fields have, since the mid-nineteenth century, succeeded in gaining increased amounts of public support for their work.

While it is now clear that American science in the late nineteenth century was not as backward as has often been assumed, there were, nevertheless, distinctive patterns evident in American research. For example, as Robert Bruce points out in his paper, the westward expansion, which constantly opened up new bonanzas of specimens, formations, and natural phenomena, did have the effect of producing a large number of scientists with more than one specialty. With so many discrete opportunities before them, American scientists tended to aim directly at one small target of opportunity after another, regarding each as an end in itself. The physical environment may indeed have tended to produce in Americans a greater emphasis on classification and description than was true of scientists generally; it may also account in part for that long-persisting allegiance to a "Baconian" inductivism that some scholars have noted. Whether the emphasis was purely a matter of geography or a more general "culture" influence cannot be determined in the absence of comparative data from other similar areas.

---

1. Don K. Price, *Government and Science* (New York, 1962), p. 27.

It is also clear that the political environment has had a pronounced effect on the development of American science. Since scientists in America have been dependent for their very existence upon external material and political support, they have necessarily been in competition with other social groups for their share of the public resources. Only when they have been able to succeed in creating alliances with members of more powerful groups—as the experiment station scientists, described in Charles Rosenberg's essay, did with members of the business, agricultural, and political communities—have scientists been able to achieve a high degree of support. Thus, geologists early in the century and, later, chemists were able to adapt to political action and thereby convert their intellectual resources into effective means of political influence. At an earlier period, geologists, acting in concert with particular agricultural, mining, or land development interests, were able to produce the necessary political "muscle" to obtain state financial support for their survey work and, in most cases, for the publication and distribution of their findings as well. A later generation of chemists, through similar means, founded agricultural experiment stations, expanded the Bureau of Chemistry of the Department of Agriculture, and made chemical analysis an important function of the Geological Survey. Physicists, on the other hand, as Daniel Kevles shows, had no such means until the twentieth century. Thus, they were unable to achieve a strong institutional base, and physics in America long remained deficient. Physicists had no organization or journal which could establish professional standards, little support from the universities, which emphasized the sciences as only a tool for producing mentally disciplined students, and no direct avenue to political power. Physicists long remained in an institutional void.

The connection between science and technology has often been debated. The general direction of thought among the new group of historians is toward the fusing of history of science and history of technology. As other works have suggested, the possibility of fusion in American culture seems to be one of the marks of its distinctiveness. Helmut Krauch, for example, has focused on the gap that exists between theoreticians, on the one hand, and practical engineers in business, on the other, in Germany. The fact that the gap did not exist in the United States, so Krauch believes, explains the accomplishment of our atomic effort.[2]

A. Hunter Dupree, offering an analysis in terms of general systems theory, suggests that both science and technology can be regarded as information systems by which man has sought to relate to some part of his environment. His concept of "measurement groups" is intended to be a way of reducing science, technology, and general culture to common terms for purposes of analysis. The nineteenth century, in his view, can be seen primarily as a period of increasing sophistication and advancing techniques of scientific and technological measurement activities. Edwin Layton considers science and technology to be "mirror-image twins" and emphasizes the importance of many individuals who held membership in both communities and were thus able to perform the extensive reformulation necessary for transmitting knowledge between the two different "information systems," to use Dupree's terminology. In Layton's view, however, the coupling of science and technology also had unfortunate consequences. In accelerating the pace of technological change, the combination resulted in greater social dislocation. It also led the engineers as a group to emulate the dispassionate, objective stance of the scientists, thus rendering them incapable of addressing themselves to the social problems they had done so much to create.

Carroll Pursell, who explores the importance of the entrepreneurial climate in the United States for the development of an interface between science and industry, emphasizes the constant entrepreneurial activity in a pluralistic setting in which all kinds of combinations of interests were possible. The scientist-consultant and engineer-manager were only two of the most evident combinations in the late 1800s. According to Pursell's persuasive argument, engineers who were trained to join theory with practice quickly moved into administrative and entrepreneurial roles in the developing corporate structure of late-nineteenth-century America. Their penetration of the industrial establishment was so pervasive that, by the end of the century, their values had become the standard of American

---

2. Helmut Krauch, "Resistance Against Analyses and Planning in Research and Development," *Management Science*, XIII (December, 1966), C–52.

society. Engineering virtues of hierarchical organization, discipline, rationality, utility, efficiency, and uniformity were generalized to become the guiding principles of the corporation and, eventually, of the federal government.

Bruce Sinclair, approaching the subject from another point of view, describes the tendency of Americans to think in terms of technical education as higher education and of the polytechnic institute to think in terms of educating men who would be managers as well as technical experts. This was both a reflection of and an important contribution to the entrepreneurial climate. Charles Rosenberg also emphasizes the importance of the research entrepreneur, the man who was able to mediate between the world of science on the one hand, and, on the other, the social and economic realities of a particular state constituency.

The contributors to this reappraisal have not answered all of the questions about nineteenth-century American science, but their essays collectively succeed in clearing away the old misunderstandings and in pointing to new directions for further research that will fill in the gaps that still exist. Further study of the social origins of American scientists, for example, as Bruce suggests, may lead to a better understanding of the conditions for the growth of science. The role of entrepreneurship in science deserves further study. Institution development and the growth of specialization, it is clear, should be major themes of new work on the period. Finally, there is nothing more evident than the great need for comparative data on science in other cultures. Once these are provided, a greater understanding of the distinctive features of American science will emerge.

I

Approaches to the History of American Science

The History of Science in America: Development and New Directions

Any examination of the current condition of and future possibilities for the study of the history of science in America will need to say something about the origins and development of this area of inquiry. This analysis is subjective and affirms that American scientific history is important insofar as it is viewed as but one entity in the spectrum of national development. The field has now reached a level of conceptualization that can be classified as professional; it is therefore important to step back and gain a sense of perspective and direction. The designation "professional" indicates a series of acceptable modes by which to approach problems, a reservoir of research materials, and a sense of identity and pride of achievement held by a body of scholars whose work is recognized as significant and competent by their peers and in the wider profession of historical studies. This volume is a signpost of the new professionalism; it is not only a unique event in the historiography of American scientific history, but it signifies that there exists a group of men familiar with one another's work, aware of common research interests, and eager to share ideas within a generally accepted style of communication and discourse.

In 1957, at the National Science Foundation Symposium on the history of science held at the University of Wisconsin, there were no papers on discussions of American topics,[1] while the 1960 Arden House conference on American civilization viewed science in Amer-

---

1. Marshall Clagett, ed., *Critical Problems in the History of Science* (Madison, Wis., 1959).

ica as at best an ancillary aspect of culture in contrast to philosophy, religion, and literature. The example of the one early effort to assess the history of American science in a national and modern scope— that of the American Philosophical Society in 1942—is instructive. The subject matter was confined to "early American science and learning," and this time period, by definition, was the single area that scholars of the 1940s equated with significance and excellence in American scientific history. The quality of the papers presented ranged from the amateurish to the semiprofessional. As a whole, they failed to provide significant insight or definition to an important subject, and only a few of them were based on manuscript sources.[2] One need only compare that experience with the conference and volume on needs and opportunities for study in the history of early American science organized in 1955 by Whitfield J. Bell and others to see the contrast between the preprofessional and the professional stage of development.[3] In 1949, moreover, Dr. Bell, in an unpublished address, analyzed the status and treatment of American science in leading American history textbooks and found it both insufficient and inaccurate.[4] Generalists are now at least able to draw upon a rich body of biographical and monographic literature that dispels the mythology of equating science with technology, assesses the role of key individuals in the history of science in America with accuracy and breadth of view, and places the relationship of science to American social thought, institutions of government, and higher education in a new and revealing light.

THE CONTEMPORARY SCENE

To offer an analogy with the past, the newly gained status of this area of inquiry is analogous to the achievements of the "Scientific

---

2. "The Early History of Science and Learning in America," *Proceedings of the American Philosophical Society*, LXXXVI (1943), 1–119.

3. Whitfield J. Bell, *Early American Science: Needs and Opportunities for Study* (Williamsburg, Va., 1955).

4. Whitfield J. Bell, "Science, Medicine, and Technology in College American History Texts" (Paper delivered at the 1949 meeting of the Mississippi Valley Historical Association). Dr. Bell, in personal conversation, indicated that he felt the situation, if reexamined in terms of contemporary textbooks, would not be greatly improved.

Lazzaroni." These were a group of powerful and centrally located decision-makers of mid-nineteenth-century American science and culture that included such men as Louis Agassiz, Alexander Dallas Bache, Joseph Henry, Benjamin Peirce, and James Dwight Dana, and often employed the aid of key politicians and nonscientists. Their primary achievement was the establishment of the National Academy of Sciences in 1863. From their beginnings in the early 1850s until the late 1860s, the Lazzaroni had certain primary goals in mind: the ridding of science of charlatanism, quackery, or what they liked to call "old fogeyism"; the establishment in the minds of the general public, and government, university, and institutional decision-makers of the idea that science was to be evaluated by those most competent to do so, namely, professionals engaged in its advance; and finally, the establishment of institutions of learning and the funding of scientific research and education comparable to that enjoyed by European *savants*. While sometimes failing in specific efforts, they nevertheless succeeded in making accommodations between expertism and populism in American democratic society and in laying the professional basis for the status of science in the twentieth century. The point of the analogy is, simply, that the study of science in America is now at a clear level of identifiable achievement; the discipline has reached the point where the problems and disputes of the past are no longer meaningful. But, as with the case of the Lazzaroni, it is highly instructive to analyze the ways in which this professionalism developed in terms of conceptualization, central figures, and prospects for future evolution.[5]

Since the mid-1940s, the study of the history of science in America has succeeded in achieving status comparable to two older historiographic traditions, namely, general American history and the history of science. It has also come to enjoy an integral and integrated

---

5. I am in disagreement with the conclusions reached by Mark Beach in his article on the Lazzaroni in this volume. I refer the reader to relevant passages in Edward Lurie, *Louis Agassiz: A Life in Science* (Chicago, 1960); *idem*, "American Scholarship: A Subjective Interpretation of Nineteenth-Century Cultural History," in *History and Literature*, ed. Robert H. Bremner (Columbus, Ohio, 1966), pp. 31–80; Richard J. Storr, *The Beginnings of Graduate Education in America* (Chicago, 1953), pp. 82–93; and Nathan Reingold, "Alexander Dallas Bache: Science and Technology in the American Idiom," *Technology and Culture*, XI (April, 1970), 163–77, especially n. 1, p. 163.

status with that conglomerate of academic and popular specializations called "American studies," whose rise to professional prominence it paralleled. In the evolution of the study of American science, there can be seen an encapsulated history of the larger pattern of historical studies in America, a scholarship which became increasingly broad in scope and increasingly integrated with humanistic, social scientific, and even scientific disciplines that had previously been held at arm's length by the historical establishment of pre-1946 days. Moreover, Lee Benson's view that historical studies need to be more conceptual and oriented toward "understanding the past in order to control the present and the future" is apt and important for future directions in the area of scholarship under consideration here.[6]

But for the time being, at least, men who work in this field can enjoy, as never before, identity and status both within their own academic halls and in the larger professional world. The field boasts a body of scholars who are continuing to produce general and special studies that characterize the course of scientific development both internally and in terms of wider patterns of social and intellectual history. A modest position in academic organization is represented by a small but growing body of text materials, a place in the curriculum structure of undergraduate and graduate education, the training of new professionals by older ones, and the correlative identification, codification, and utilization of a body of research materials. These materials, formerly difficult to obtain and rarely utilized, have been shown to throw new and unexpected light on the careers of many scientists of various degrees of renown. These resources range from serial publications to oral histories to manuscript resources to the published and unpublished works of the men themselves. The histories of education, of the relationship of government to society and ideas, of scientific techniques, and of popular attitudes and values are only a few of the areas of study made sophisticated by scholars whose primary interest was science in the American context. Understandably, the training of advanced students has been made easier by these research resources, and the products of these students' labors

6. See *Daedalus*, XCVIII (Fall, 1969), 801–86.

6

have gained more ready acceptance within the established academic community because of the responsibility of their mentors and the broad-gauged nature of the data on which the conclusions of research have been grounded. One needs, for example, only to compare such works of the preprofessional period as Dirk J. Struik's *Yankee Science in the Making,* Merle Odgers' biography of Bache, or Fulton and Thompson's biography of Silliman with such products of the professional years as William Stanton's *The Leopard's Spots,* Brooke Hindle's *Pursuit of Science,* or Nathan Reingold's *Science in Nineteenth-Century America* to see the distinction between older and newer modes of research that signify newly achieved professionalism.[7] The works of the 1940s, important in terms of their groundbreaking and innovative character, were by and large written by interested nonhistorians, who relied on printed materials or on one readily available set of manuscript materials.

One immediate result of the increased stature of the field has been the relative ease with which young men who seek out and exploit these materials are able to gain acceptance in the academic community through university employment, an achievement far superior to the conditions of twenty or even ten years ago. Moreover, the increased hiring of such individuals at major colleges and universities increases the opportunity of teaching advanced students in their own specialty. Added to this is the fact that many special and technical schools that teach biomedical, physical, or technological sciences have seen the virtue of the study of these disciplines in the American context. Obviously, the situation is still not as rosy as that recently described by a traditional historian of science in the *New York Times* when he affirmed that his specialty had become so notable in the academic pecking order that any man worth his salt

---

7. Dirk J. Struik, *Yankee Science in the Making* (Boston, 1948; rev. ed., New York, 1962); Merle M. Odgers, *Alexander Dallas Bache, Scientist and Educator: 1806–1867* (Philadelphia, Pa., 1947); John F. Fulton and Elizabeth H. Thompson, *Benjamin Silliman, 1779–1864: Pathfinder in American Science* (New York, 1947); William Stanton, *The Leopard's Spots: Scientific Attitudes toward Race in America, 1815–1859* (Chicago, 1960); Brooke Hindle, *The Pursuit of Science in Revolutionary America, 1735–1789* (Chapel Hill, N.C., 1956). See also *idem, David Rittenhouse* (Princeton, N.J., 1964); Nathan Reingold, ed., *Science in Nineteenth-Century America: A Documentary History* (New York, 1964).

could, until a few years back, increase his salary 50 per cent by moving from one position to another.[8] That this situation does not pertain in American science signifies that specialists in this area have generally been content to identify their roles and goals within the broad spectrum of American intellectual or cultural history rather than in the history of science as a separate discipline.

In terms of the sociology of professionalism, the study of the history of science in America has come full circle. Its beginnings were in the liberal and innovative imagination of American historians who shaped the character of the profession in the immediate postwar years and influenced their students to move away from the norms of political, economic, and diplomatic history as standards of research and employment. Thus launched, these young scholars experienced some obstacles and academic misunderstanding in uncovering research materials, gaining the cooperation of scientists, and competing with other professionals such as the historians of science. By the late 1960s, however, new people who came into the field had come to enjoy fully the status that their intellectual and academic grandfathers had confidently expected.

It is a happy fact of organizational existence that the history of science in America can be celebrated yearly on panels held at the American Historical Association, the Organization of American Historians, the American Studies Association, the History of Science Society, or the Society for the History of Technology. Sociologists pay careful attention to the role and function of science as a historically grounded institution in America, and political scientists are increasingly receptive to the uses of history in explaining the relationship of science to governmental processes. For psychologists and other social scientists, the identification of personality types in science is becoming increasingly important.

Contemporary students of science in the American context are beginning to enjoy a more pleasant institutional universe. They have been accepted, if sometimes grudgingly, by traditional historians of science. Such acceptance is still partially contingent on their study-

---

8. Robert Reinhold, "Rise of Scientific History is a Reply to Technology," *New York Times*, February 18, 1970, pt. II, pp. 1, 51.

ing what is sometimes referred to as science that "contributed" to the internal advance of a special area, comparable to studies done by traditionalists who specialize in the history of science in Europe. It is even acknowledged by them that the history of science—in America and elsewhere—has a "social aspect" even though this aspect is rarely defined or treated with precision. Hence, as full-fledged students of science in America, younger scholars do not need to feel a dualism of identity as either students of American thought and culture or historians of science, a professional condition that often characterized their teachers. If research support is required, this is now available from a number of public and private funding agencies, and the titles of projects so supported in recent years clearly indicate a shift in regard concerning the merits of studying the history of science in America.[9]

THE PREPROFESSIONAL YEARS

In the march toward professionalism, several books served as signposts along the way. There are important prefigurations in the ideas and data contained in the work of Arthur M. Schlesinger, Allan Nevins, Bert James Loewenberg, and Henry Steele Commager. Richard Hofstadter's *Social Darwinism in American Thought* (1944) is a work typical of the early thrust toward broadening the horizons of American history; it demonstrates how science, viewed broadly as part of the social context and the thought pattern of an era, can provide a key to understanding that is more sophisticated and broadly based than reliance on the history of religion, politics, or "society" as loosely conceived by an earlier generation. One can see the same kind of emphasis in Allan Nevins' chapters on ideas and culture in *Ordeal of the Union*.[10] This focus is also evident in

---

9. See Edward Lurie, "An Interpretation of Science in the Nineteenth Century: A Study in History and Historiography," *Journal of World History,* VIII (1965), 681–706. See also the implicit assumptions regarding methodology in the history of science revealed in Clagett, *Critical Problems.*

10. Arthur M. Schlesinger, *The Rise of the City, 1878–1898* (New York, 1933); see also *idem,* "An American Historian Looks at Science and Technology," *Isis,* XXXVI (October, 1946), 162–66; A. Hunter Dupree, "The History of American Science—A Field Finds Itself," *American Historical Review,* LXXI (April, 1966), 863–74; Allan Nevins, *The Ordeal of the Union,* 2 vols. (New York,

the concept of relating science to history, politics, law, psychology, sociology, theology, and literature demonstrable in Stow Persons' *Evolutionary Thought in America.*[11]

When historians of science in America began to study both the patterns and the specific aspects of development, the picture that emerged sometimes differed from the one drawn by those pioneers who first generalized about the role of science in society. The story of the evolution controversy is, of course, a classic case in point. Hofstadter, building on the work of precursors of the 1930s, made assumptions about the nature of the clash of ideas and systems of belief that were restatements of the older concept of "warfare" between religion and science, an idea that took a long time to die in general and scientific historiography. The basic difference between the early enthusiasts of the science-in-society concept and the later, professionally oriented group resided in the ability of the latter to cut through old assumptions about the significance of science because of their thorough familiarity with a range of manuscript materials.

In the works of the new group of professional historians of science in America attention to detail is of high importance, as is the desire to characterize and define the exact extent and nature of a subject's importance to science. Here there occurs a fairly sharp break with the older tradition, a break that is the result not merely of diligence and the desire to lay to rest old myths but of the conscious effort and felt need of the new professional students to prove their worth. This sense of value and identity has had to be established within the two camps that were the temporary homes of those who established the history of science in America as a respectable area of inquiry.

The two sources of identification for early historians of science in America were the history of science and general American history.

1947), especially those sections on cultural history dealing with science. On Loewenberg, see his pathmarking article, Bert James Loewenberg, "The Reaction of American Scientists to Darwinism," *American Historical Review*, XXXVIII (July, 1933), 687–701; Richard Hofstadter, *Social Darwinism in American Thought*, rev. ed. (New York, 1959). See also *idem*, *Anti-Intellectualism in American Life* (New York, 1963), especially pts. III, V, and VI.

11. Stow Persons, ed., *Evolutionary Thought in America* (New Haven, Conn., 1950).

In the early 1950s, the only established spokesmen for American science were the mathematician Dirk J. Struik, the sociologist Bernard Barber,[12] and the historian of science I. Bernard Cohen. It seemed that the new historian of American science would be content to continue on the path already partly cleared by the pioneers of social or intellectual history who made their mark in the 1930s. But, in building on the enthusiasm and encouragement of men like Merle Curti, the late Richard H. Shryock, and Arthur M. Schlesinger, the new man found himself, in the early 1950s, facing a research and professional situation far different from that of the historian of ideas and culture whose work had been done in the years prior to the Second World War.[13]

THE BEGINNING OF PROFESSIONALISM

The new condition can be described as the search for identity, validity, and independence of viewpoint. The problem of identity was perhaps the most important. Unlike their mentors, who were content to take a secondary source at face value or to proffer an idea or insight to be developed by someone else, the historians of science in America of the postwar years had to do primary research to establish their competence. This need was congruent with the necessity of gaining the support of scientists, general historians of ideas of culture, and the newly respectable historians of science. Coming from a tradition of social and intellectual history, the historian of science in America found it difficult to justify his existence to those who insisted that he had to have worked in science in order to understand it, or to the historians of science and others who implied that the American-oriented endeavor was of secondary importance. Whereas the teachers of these postwar historians had begun as re-

---

12. Bernard Barber, *Science and the Social Order* (Glencoe, Ill., 1950).

13. See Merle Curti, *The Growth of American Thought* (1943; rev. ed., New York, 1964); Schlesinger, *The Rise of the City*; and, for the stellar contributions of the late Richard H. Shryock to the understanding of the social aspects of science, see Shryock, "The Need for Studies in the History of American Science," *Isis*, XXXV (April, 1944), 10–13; see also his prescient Anson G. Phelps Lectures at New York University (1959), published as *Medicine and Society in America, 1660–1860* (New York, 1960).

spectable American historians and expected their students to master and eventually teach in that area, they sent the students out into a world where they would be evaluated by those who were affiliated directly with the scientific community or who, like the majority of historians of science, stemmed from a scientific background. Most historical mentors, conversely, had done their first work in political, social, or intellectual history, and had only developed an enthusiasm for science in the latter part of their career. Even those singular men like Richard H. Shryock, who had successfully bridged both worlds, represented a fundamental interest in the social aspects of the scientific disciplines. In encouraging students to branch out from the traditional specializations in economic, political, or diplomatic history, these early mentors had not anticipated the doubt of scientists and the condescension of historians of science that met their students. This could hardly have appeared much of a problem at the time, because historians of the prewar years had seemingly made perfectly acceptable and respectable judgments on religion, literature, law, and politics, and it seemed equally reasonable that their students could examine physics, zoology, or technology in the same manner and with the same proficiency.

But if this entrance into a domain heretofore regarded as the popular property of anyone who had access to an ancestor's papers or who was interested in the influence of science on the social order did not seem to present much of a problem to teachers, it was at times an uncomfortable condition for their students. These new scholars sometimes found themselves viewed with suspicion by scientists, who grew up in the older tradition of writing their own history and now found it curious and somewhat unbelievable that a nonspecialist could generalize on the significance of the work of their intellectual forebears. This is not to say that established scientists were obstacles to research; they were usually sympathetic, supportive, and cooperative in supplying the fledgling professional with access to rich and vast primary source materials.[14]

14. In this connection I should like to acknowledge my personal indebtedness to Alfred S. Romer, Director Emeritus of the Museum of Comparative Zoology at Harvard College, for his kindness and generosity—personal and intellectual—in providing me with insights into the history of evolution and with data on the Louis Agassiz period.

The matter of ability and evaluation was of concern both to scientists and to the new generation of historians seeking answers to new questions about the life of the scientific mind in America. Interested in examining ideas in terms of content and context, historians were initially unable to cope fully with the task of analyzing a mass of manuscript and other primary resources of research that dealt with such areas as zoology, medicine, thermodynamics, paleontology, or physics. Often, this resulted in a less than optimum utilization of these newly available, but sometimes difficult to comprehend, materials of scholarship. Scholars who wanted and needed to continue their primary identification as historians of ideas and culture had to engage in the difficult and arduous task of establishing levels of significance and competence to guide their research.[15] However, as they began to work, they learned the sciences whose past they wrote about and found them comprehensible. More than this, an awareness developed that the mind of a man could be representative of both his contribution to a specialty and his role in the general culture, and the volumes that grew out of this impulse had of necessity to illustrate this realization of varying levels of significance.[16] Historians were, after all, just that; they were not zoologists or physicists. They learned that while they could exhibit sufficient competence in describing the mechanisms and the conceptualizations that men used in thinking about nature, what was far more important was their ability to integrate this knowledge with the range of social and cultural attitudes revealed in coexistent patterns of American religion, politics, or economic and social development. This developing ability had much to do with convincing scientists to cooperate with historians in the discovery of new vistas of intellectual and cultural history, while at the same time it reassured the historians that their work represented competent appraisals of the internal character of the science being studied.

---

15. See "Conference on Science Manuscripts," *Isis*, LIII (March, 1962), 3–157.

16. See, for example, these works by students of American culture that illustrate this condition: Ernest Samuels, *The Young Henry Adams* (Cambridge, Mass., 1948); *idem, Henry Adams: The Middle Years* (Cambridge, Mass., 1958); *idem, Henry Adams: The Major Phase* (Cambridge, Mass., 1964); Wallace Stegner, *Beyond the Hundredth Meridian: John Wesley Powell and the Second Opening of the West* (Boston, 1954); Thurman Wilkins, *Clarence King* (New York, 1958).

It was perhaps more difficult for those interested in science in America to gain acceptance from general historians of science, and this was a significant need in the early 1950s. Coming from a scientific background, most historians of science enjoyed the respect of scientists and the financial support of funding agencies as well as an academic respectability in the educational marketplace that was usually unavailable to the new historians of science in America. Moreover, the decision-makers of historical studies often asked young scholars to teach traditional courses in the history of science when they were in fact unprepared to do so. Finally, the rapid rise to public and academic prominence of the general history of science profession meant that scholars of this order had easy access to a professional journal (*Isis*) and a scholarly society (the History of Science Society), an advantage not readily available to the new historians of science in America, who still had to work within the leading historical associations.

Historians of science in America found themselves forced to defend the significance of their area of study. Most historians of science, properly designating science as international or supranational, could not quite understand the meaning of an "American" science. They did credit the existence of a "French" science, because its leading figures were scientists of a high caliber who worked within a tradition that had been established and identified.[17] This was almost impossible to credit for the American scene. In the eyes of specialists, most of whom had done their work in the physical sciences or in history outside the American context, science in the United States was a viable area of inquiry only insofar as it was comparable to that which existed in other times and other places. While understandable in terms of the background and values of the historians of science, this evaluation made little sense to American students, who had never been forced to defend the significance of Jefferson, Edwards, Emerson, Thoreau, or Dewey. The assumption that ideas in science had to be studied in and for themselves guided the thinking of many historians of science but was usually foreign

---

17. See the implicit assumptions in the articles and commentaries in Clagett, *Critical Problems.*

to students of the American scientific environment.[18] The full result of these conditions was a curious kind of acceptance by historians of science, who often suspected the validity of the American enterprise, but, nevertheless, entertained the presence of scholars in the field because of the soundness of their work and the degree to which their social interpretation of science was valid. This accommodation was useful because, if nothing else, it made historians of science in America more precise and rigorous in their scholarship. It also produced a significant examination, within each group, of the validity of time and place.

It was ironic that the degree to which the historians of science in America gained acceptance from scientists or historians of science was sometimes a measure of a growing distance from that very group—historians of American society, ideas, and culture—that had nurtured them. With few exceptions, the individuals who worked in American science did not achieve status as major spokesmen for the general field of cultural and intellectual history. A distinction arose between those who worked in such fields as social history, religion, historiography, or ethnic studies, and those who studied American science. The reasons for this distinction are subtle, but there are some general conclusions to be drawn from this condition of the mid-1950s and the 1960s. Analysts of the scientific scene had to prove their identity as proper historians to traditionalists whose fields were politics, economics, or diplomacy, and who, by and large, made decisions with regard to professional rewards and placement. To such men, the historian of science in America, while interesting, was really a strange type. Was he a historian of science? If so, where was his scientific training? Was he expert in social and intellectual history? If so, he seemed to work too hard to establish the obvious, while the discovery and use of his research materials seemed to require more zeal and energy than were merited in terms of the results they yielded. For such men, it was difficult, but possible, to grant a place to social or intellectual analysis; the former had been established since the 1930s, while the latter, if still mysterious, had bases of reference in such questions as the clash between science and re-

18. See Lurie, "An Interpretation of Science."

ligion, the role of the Enlightenment in shaping man's ideas of "Nature and Nature's God," or the uses of intellectual history in understanding progressive thought, New Deal pragmatism, or the philosophy of reform in the Age of the Common Man.

It is not surprising, then, that the scholar in the American field who wanted to trace the development of a scientific idea should find himself ill at ease with older scholars whose concepts of intellectual history were at once more instrumental, less sophisticated, and more immediately identifiable in the agreed upon social context of American historiography. When such an individual preferred to work on heretofore unexplored problems in the relationship of scientists to their environment (e.g., the changing significance of science in higher education, or the relationship of science to various aspects of romanticism in literature or social theory) the result was usually limited acceptance. Perhaps there was a thread of antiscientism woven into the evaluative processes of men who felt pressured, both in academic environments and in the world at large, by the increasing social domination of science. Perhaps, too, there was impatience with a lack of the kind of overview of science that had been provided by the original inspirers of science-oriented social and intellectual history.[19] Whatever the cause, this reaction formed a general pattern in what may be termed the first period of professionalism, the years from about 1945 to 1960, when students of the American field were facing three sets of obstacles: the discovery and definition of materials for research, the achievement of accommodation with the profession of the history of science, and full acceptance by their fellow American historians.

The historian of science in America was not a lowly misanthrope who was despised by most classes of academics, nor was he faced constantly with the necessity of rationalizing and justifying his intellectual position. Many historians of American science in this period enjoyed handsome research support, happy academic homes,

---

19. See citations above to Curti, Loewenberg, Nevins, Schlesinger, and Shryock. To this group should be added Howard Mumford Jones, who, in the selection of reprinted and edited titles during his tenure as editor of the John Harvard Library, is notable for the attention he gave to reprinting classics in the history of science in America.

and acceptance in all the camps described; however, in this formative period, the historian of the American scientific environment often had to engage in unnecessary and wasteful disputes and expenditures of energy to gain the respectability that was due him. The very magnitude of the tasks, both professional and intellectual, before him meant that he rarely had the opportunity to sit back and take stock of where he and his relatively small band of companions had been and where they should be going if their area of expertise was to be of utility to the study of America and to the understanding and control of the present and future. An exceptional opportunity was the 1962 Conference on Science Manuscripts, organized by Nathan Reingold; another was an entire issue of the *Journal of World History* in 1965 devoted to assessing the nature and role of science in the American context.[20]

THE EMERGENCE OF A NEW PROFESSIONALISM

One basic reason why scholars of science in America have by now achieved confidence, acceptance, and identity as notable American historians has been the overall high quality of their recent work. If the generalist who writes college texts were to integrate the findings about natural history, the physical sciences, and the institutions of science, and the studies of the eighteenth and nineteenth centuries that have come from the pen of specialists since the war years, the record would be most impressive. The essays in this volume are testimony to this. It is this impressive performance, with the high level of work typical of the students of the first professionals, that marks the history of science in America as an area of inquiry that has come of age.

The next phase of professional development needs careful guidance and conceptualization. Books on methods or the philosophy of the field have been rare. Part of the reason for the absence of overall efforts at conceptual emphasis resides in the problem of duality of identity already discussed, as well as the need to produce high-

---

20. "Conference on Science Manuscripts"; "Science in the American Context," *Journal of World History*, VIII (1965).

quality monographs that will give the field its due respect. As there has been a failure to think of American science in terms of overarching concepts that define and characterize its development, so too has there been little effort to fill in the evident gaps in mere information that might provide the underpinning for such a general conceptualization of the significance of the field.

DIRECTIONS FOR THE FUTURE

The following conceptual and descriptive areas of research may be viable tasks for the immediate future. Of prime importance is an overview of the history of science in America. With the amount of spadework already accomplished and the research materials available both in private collections and government holdings, it should now be feasible to gain a comprehensive view of the discipline. It might be possible to write a general, but valuable, history of nineteenth-century American science using no other sources than those so finely detailed and placed in perspective by Nathan Reingold in his collection of documents. It will be a long while before all the men listed by Whitfield J. Bell in *Early American Science* have received their scholarly due; properly conceived and diligently researched biographies of Joseph Henry, Josiah Willard Gibbs, or Spencer F. Baird are but some of the studies needed to complete professional knowledge of the nineteenth century. Histories of modern genetics and physics still stand as open spaces on library shelves.

In terms of conceptual history, future work could emphasize what science is and what it is not, by separating out the actual bases of American excellence from popular belief about such virtue. It is important to understand what a man like Joseph Leidy or James Dwight Dana meant when he used the concept of science in relation to that aspect of nature he studied. A scientific subject could be interwoven with public conceptions of it (as John Greene did for the "age of Jefferson") [21] and could also be balanced against idealized views and preconceptions that the scientist held about himself and

---

21. John C. Greene, "Science and the Public in the Age of Jefferson," *Isis,* XLIX (March, 1958), 13–25.

18

his profession. This kind of study, for example, might do much to explain the real significance of the Scientific Lazzaroni, viewed both as contributors to knowledge and as men who prefigured the course of the future. One could chart the emergence of professionalism as a quality central to aspects of scientific existence—from its late eighteenth-century beginnings, through its dilution in the heady days of the republican deification of nature and its recapture by the patrician directors of culture and the mugwumps of the years from the pre-Civil War era through the 1880s, to the emergence of the modern spirit (it too cyclical) that dignifies research and study in the years from the late nineteenth century onward.[22] The role of science in the federal establishment, in the structure of higher education, and in the politics of establishing priorities and decision-making are all aspects of historical evolutions that such an overview could consider.

Also important would be a general history of science in America which actually makes direct correlations between scientific endeavor and other, equally notable components of national values and cultural and intellectual history. It is also apparent that a modern synthesis of American scientific history should have as a central concern the theme of the duality of elitist ideas and democratic popular thought—a theme that is to be found deeply embedded in the unconscious and conscious activities and value systems of scientists, as well as those of the various publics with whom they are conjoined. The history of societies and associations of scientists, the relationship of scientists to other decision-makers, the attitude of nonscientific shapers of opinion and creative men toward the discipline, the problem of status, the question of the control and governance of science within the boundaries of democracy—all are tied to the analysis of this elite-populist dichotomy.

Finally, an overview might seek to assess distinctions and relationships between science in America and science elsewhere. Despite the title of this paper, and the use of the term *science in America*, it is not yet philosophically established whether *American*

---

22. Edward Lurie, "Science in American Thought," *Journal of World History,* VIII (1965), 638–55.

science would not be a better descriptive term, because the question of whether science is international or is influenced primarily by its environment has not yet been treated in any systematic or thoughtfully contributive manner.[23]

The nineteenth century provides the historian with excellent case material for the testing of conceptual hypotheses. The Lazzaroni represented a personality type seeking to associate American endeavor with larger European patterns from which they felt estranged. Is it then possible to identify aggregations of their opponents as symbolizing a type of scientific nativism? If the efforts of the Lazzaroni were really not notable, then it is important to identify those areas of cultural decision-making and those men of power within science who blocked such efforts. Here, there is a need for studies of Spencer F. Baird, Joseph Leidy, and Alexander Agassiz as men both affected by and independent of the Lazzaroni influence. A biography of Alexander Dallas Bache, together with a history of the Smithsonian Institution and the National Academy, would yield riches for the study of the elitist spirit. If that spirit can be identified as continuing into the twentieth century, what does the persistence of this impulse imply about the seeming maturity of American men of science with regard to European endeavor? If the Lazzaroni spirit is still with us, and forms the essential linkage between secret science and secret government, then it would seem that scientists are inevitably in a state of continued tension and accommodation at odds with the public of their time.

The case of the Lazzaroni suggests the importance of group biography. There are other cases in which this technique would be of value. Similarly, it would be important to identify the characteristics of regional science in mid-nineteenth-century America as typified by the scientific circles in Boston, Washington, and Philadelphia. A monograph on the rise and fall of science in Philadelphia would not

---

23. If the American environment can be evaluated as having exerted singular influence upon intellectuals and ideas so as to make the life of the mind qualitatively different on these shores, the place of science in American civilization will be furnished with a means to evaluate this concept. I suspect that the answer resides in the different qualities of the time periods under study. In this context, see Daniel J. Boorstin, *The Americans: The Colonial Experience* (New York, 1958).

only tell us much about the origins of the Lazzaroni, but would also reveal why men raised in the spirit of Franklin were prominent only insofar as they ceased to be provincial and became figures on the national scene. The technique of group biography could be pursued fruitfully in this context, and the historian of the nineteenth century in America could identify such men as Henry, Baird, and Bache as notable personalities who, existing in the same locale, influenced ideas, action, and social roles. The same sort of study could focus on the Le Conte family, whose intertwined careers were indicative of the natural history in the South, the West, and the Middle Atlantic states. A collective study of the Rogers family would illuminate the character of the natural history and the physical sciences in the South, in New England, and in the Middle West.

If Joseph Henry's career is a reflection of much of the character of nineteenth-century science, a comparison of his life with that of a man such as James Hall would be revealing because Henry operated as a national figure, while Hall, an eminent paleontologist, was content to spend his life in Albany, New York. In addition, the institutional history of science in nineteenth-century America remains to be written. We know hardly anything of its development in such central university centers as New York, Boston, Cambridge, and Philadelphia. The role of science in the development of American higher education still awaits its historian.

New biographies could supplement past research. Life histories of some major figures in the evolution controversy—Jeffries Wyman and William B. Rogers, for example—are still needed. These are but some of the conceptual and informational problems that still await the skill of the professional historian of science in America. The study of science in America has now reached a mature level; its new generation of scholars must now ask the right questions and carry on studies in both history and theory that will advance the discipline to the status inherent in its content. In these ways, the early vision of the first inspirers of this vital subject will hopefully be realized.

The Measuring Behavior of Americans

Over the past twenty-five years, the history of science and technology in America has attracted a number of scholars who have built a literature and have worked out in rough form both a scenario and a *dramatis personae*.[1] Despite these accomplishments, however, little progress has followed either in establishing generalizations about the field or in tying them to the mainstream themes in history, in the history of science, or in science. Students of the history of American science have customarily paid lip service to the proposition that science is a social activity, but they have derived few convincing or very profound arguments from the general statement. They have, by linking science and nationalism, brought on themselves the suspicion of chauvinism. They have not countered the charge that American science is derivative and unoriginal. They have not credibly established a connection *or* a separation between science and technology. They have not countered the charge that the social history of science is "external" to science and that what is scientific, i.e., "internal," must escape their method. They have not developed a credible theory of the relation of applied science to social change. They have neither defined convincingly that entity "pure science" (or "basic research," or "fundamental science") nor accounted for the persistence of the concept without a definition.

---

1. Cf. A. Hunter Dupree, "The History of American Science—A Field Finds Itself," *American Historical Review*, LXXI (April, 1966), 863–74; Brooke Hindle, *Technology in Early America* (Durham, N.C., 1966), a thorough treatment of the history of American technology.

Most important, they have not adequately tied their scenario for science or for technology to the great tides of interaction between men and their environment in North America as seen by generations of American historians who have written of economics, politics, institutions, culture, thought, and values.

PARALLELS BETWEEN HISTORY AND CONTEMPORARY POLICY

These shortcomings in judgment are strikingly similar to those in discussions of contemporary science policy. If we had well-established and strongly humanistic views on science and nationalism, science and technology, basic research and applied science, technology and social change, and the assignment of value-relating research to the environment, we could do much more than merely write more adequate histories of science and technology in America. We might help the National Science Foundation adapt its ideal of basic research both to technological applications of science and to the social sciences. We might assist Congress in designing a mechanism for technology assessment. We might be able to reorganize the government-university partnership to deal more adequately with environmental problems and, at the same time, to meet its responsibility to basic research. We might readily apply a deeper understanding of the relation of science and technology to social change. We might design ways of introducing increased productivity and positive population control in developing societies without destroying their traditional cultures. One might justifiably conclude, therefore, that a solution to current policy problems of science and technology and a deeper understanding of our history must come simultaneously rather than consecutively.

The tools of social and economic analysis which have become current in the post-World War II period have not yielded rich results when applied to science and technology. Science especially seems to escape the clutches of sociology. Peter Laslett points out that the scientific revolution was a product of "the world we have lost," the traditional society which existed before the industrial age. He confesses his inability to apply his statistical approach to science.

Scientific activity . . . during the heroic generations when Englishmen were so prominent in it, was heterogeneous, so various that it can scarcely be called an activity in itself. It has to be sought in the tiny interstices, the nooks and crannies of the social structural whole, for the number of people able to contribute to it was tiny even on the scale of their own small society, and their activities were remote from each other. The study of an activity of this sort is a study of what might be called residues, minutiae, not of a general preoccupation which can be easily associated with the trend of overall, widespread social change.[2]

SYSTEMS THEORY AND THE HISTORY OF SCIENCE AND TECHNOLOGY

While the historians have been waiting in vain for sociology to give them a social history of science, the contemporary world of policy has seen the rise of another, more comprehensive, tool of analysis—general systems theory. As a conceptual tool, it is broader than computers or systems analysis and has crept into other fields of scholarship in a welter of confused and conflicting vocabularies. Systems theory, once the property of the military planners, has surfaced in such disparate fields as anthropology and business management.[3] The concept of a systems approach, along with the concept of ecological balance, has a place in revealing the history of technology as that part of culture which patterns the behavior providing the life support systems for man.[4] One recent broad definition of technology which has had the support of such scholars as Emmanuel Mesthene and Peter Drucker is that technology is an information system which connects the species of biological organisms *Homo sapiens* with its environment. Thus all individuals, however unskilled, partake of the technological information system if they are sufficiently adapted to be alive. All human societies have a body

---

2. Peter Laslett, *The World We Have Lost: England before the Industrial Age* (New York, 1965), pp. 168–69.

3. Ludwig von Betanlanffy, *General Systems Theory: Foundations, Development, Applications* (New York, 1968).

4. A. Hunter Dupree, "Comment: The Role of Technology in Society and the Need for Historical Perspective," *Technology and Culture*, X (1969), 528–34.

of technological information as a part of their culture, and an exceedingly important though not exclusive means of storing and transporting this information is by language.

Science can be described in the same idiom:

> Science, like technology, is an information system embedded in culture. It too mediates between man and his environment. It too is a social process concerned with a memory bank which stores information and passes it from one individual to another, including those in the younger generation who will take their places in an unbroken chain. It too has embedded in its tradition a mathematics tied to a measuring system.[5]

One might add the caution that the mathematics and the measuring system just referred to have representation not only in number symbols but also in number words. In addition, behavior carries mathematical relations through tools and neuromuscular coordination. The template by which the smith hammers metal to a certain shape and thickness is no less a measurement for its nonverbal form.[6]

In this scheme, the difference between science and technology can be seen in the notion that technology must somehow either directly or indirectly contribute to the survival of the species. Science collects information from the environment but without the expectation on the part of society that adaptive behavior must be immediately forthcoming;[7] science, in the language of general systems theory, is not a closed-loop feedback system. The process of abstraction, which the scientific information system began to develop in ancient times, broke the loop and relieved the system of the necessity of producing an unbroken series of adapted behavior patterns which would suit the species to its ecological niche. The time span available for the processing of information within the system is thus greatly and indefinitely increased, and the number of optional solutions is also greater.

---

5. *Ibid.,* p. 532.

6. I am indebted for the term "template" to James Deetz, *Invitation to Archeology* (Garden City, N.Y., 1967), p. 45.

7. Dupree, "The Role of Technology," p. 533.

MEASUREMENT IN THE UNITED STATES

To apply this concept to the United States at the beginning of the Republic we need some kind of unit of measurement which will be present (1) in the general population, (2) in the groups which perform specialized technological roles, and (3) among American scientists. Inherent in the previous discussion is the idea that measurement is the activity by which men make information about the environment transmissible and comparable. In providing an index to measuring behavior, the artifacts shaped by the past measuring systems give the historian a glimpse into otherwise unrecorded segments of the history of science and technology. Measuring behavior (which should be distinguished from the narrow range of subjects treated in an antiquarian and authoritarian way by experts on weights and measures at bureaus of standards) is something that all men do, but the form measurements take is a mark of the total culture of a people and a way of distinguishing one culture from another. Among restricted groups measurement also plays a part in defining the distinctive information possessed by those groups. In technology a craft guild preserves its measuring technique in its mystery. Entrance to a craft is determined by an apprentice's ability to perform certain measurement activities as compared with the expertise of the masters of the guild. The measuring activity called science has much in common with the craft but differs in its explicit concern with the processing of information according to its own time schedule.

In late-eighteenth-century America, a rich matrix of measuring behavior already existed in all three groups we are considering—the total population, the crafts, the sciences. The total population was not limited to individuals of European origin. The Indian had a patterned measuring behavior, which he later meshed with European culture—as in the adoption of firearms.[8] A perfectly legitimate study might be to look for evidence of African patterns of measuring behavior among the black immigrants and their descendants, in-

8. I am indebted to Patrick Malone for information on this subject.

cluding those craftsman slaves about whom Frederick Law Olmsted made such persistent inquiries.

The European immigrant brought with him or her types of measuring behavior drawn from medieval and Renaissance science and technology, which he continued as he adapted to the New World.[9] This adaptation to the new environment forced the farmers and craftsmen who made up the European population of America to more intensive measurement than had been necessary in seventeenth-century Europe. In their efforts to measure their own juncture with the environment, i.e., land, they had more in common with the Benedictines and Cistercians of earlier centuries than with their own European contemporaries.

Despite the absence of formal guilds in colonial America, the heritage of the new Republic included a full set of highly developed crafts in the Western European tradition. A craft had its mystery, largely centered around its measuring activity, by which behavioral templates that passed from master to apprentice were impressed upon the environment. These templates were not available to the whole population but only to those who had been initiated into the mystery by long apprenticeship. Part of the craft tradition was carried by language, but an important part was carried in the apprentice's imitation of the master's motions. Nearly all of the tradition could be preserved without a written record, and the constant matching of the work of the apprentice to the work of the master insured great stability in the craft tradition, even when the surrounding environment and the materials it provided the craft underwent marked change. Until the late nineteenth century, the best way to transport technology from one place to another was to transport artisans. The stabilities of craft technology remained intact both in crossing the ocean and over long spans of time after the crossing. A barn or a house built in New England in the seventeenth century shows the same scarfs and frame designs as those found in England.[10] A twentieth-century English expert might understand-

---

9. Lynn White, Jr., "The Legacy of the Middle Ages in the American Wild West," *Speculum*, XL (1965), 191–202.

10. Cecil A. Hewett, "Some East Anglian Prototypes for Early Timber Houses in America," *Post-Medieval Archeology* (1969), pp. 100–121.

ably date a barn in the Connecticut valley as seventeenth-century only to find that the owner had papers to prove its erection in the 1860s.

Thus, highly structured measurement activity formed the core of craft organization in the young Republic. The craft groups, even without formal guilds, were able to control standards sufficiently to keep out less skilled members of the population although basic craft knowledge became more widely diffused in America than in Europe.[11] The crafts formed the basis for both agricultural and industrial technology. They were a part of the ecological support system of the species and were recognized as such by Americans, who accordingly put a high value on practicality.

MEASUREMENT AND SCIENCE IN AMERICA

Scientific endeavor in America at the end of the colonial period was extensive and has been well documented.[12] Two paradoxes have bothered all students of late eighteenth-century science: (1) Scientific activity was closely associated with the "arts," that is, the craft traditions. Yet science was clearly a different sort of activity organized into cosmopolitan networks across the Atlantic and heavily dependent on formal education and its own literature for the transmission of information. (2) Science was not an established profession—as law and medicine were—but it was, nevertheless, a clearly defined activity not to be confused with the crafts or simply designated as general intellectual curiosity and enthusiasm.

A way around these two paradoxes is to consider science as a measurement system, the object of which is to observe some aspect of the environment and to store this information in a literature (memory bank) where it then can be processed according to the agreed-upon rules of the tradition. The scientific community is like a craft in that it has its own mystery and its own standards of admittance to the group. Its measurement system is also transmitted in part by

---

11. Carl Bridenbaugh, *The Colonial Craftsman* (New York, 1950).
12. See Brooke Hindle, *Pursuit of Science in Revolutionary America, 1735–1789* (Chapel Hill, N.C., 1956).

language and in part by demonstrated skills. Science differs from the traditional craft, however, in kind rather than by degree. Scientists claim for themselves immunity from the obligation to contribute directly to the survival of the society. In contrast to the craft guild, the scientific community has as an article of faith the proposition that the information taken in must be processed according to rules of consistency internal to the body of scientific information. The spokesmen for the scientific community are quite willing to promise —on the basis of past performance—results that will enter the survival system of the society and possibly enhance dramatically the operation of the system. At the same time these spokesmen always warn that the placing of a time limit—however far in the future—at which they are obligated to deliver a practical answer is futile. The imposition of a time limit would only inhibit the internal processes. Thus science, as distinct from a craft, serves its social function by eschewing short-run practicality. Hence the scientist argues for abstraction and high standards of internal consistency in the belief that such practices will enhance the ultimate social usefulness of scientific measurement.

One can imagine a highly structured society in which both craft groups and the scientific community were completely and exclusively professionalized. If, in addition, these professional groups were to become highly differentiated into specialized behavior patterns, an inability to move between groups and subgroups would be almost complete. Such tight professionalization and specialization did not characterize the early days of the Republic. The barriers between the arts and sciences were low enough that a Benjamin Franklin could jump over them with ease.

A good example of measuring behavior in the late eighteenth century may be found in the career of David Rittenhouse.[13] His early technological accomplishments as a clockmaker and instrument-maker were in a craft setting, but they gave him measurement abilities with which he could attack a problem of science. In addition to his instruments, his own personal skills were needed for the ob-

---

13. See Brooke Hindle, *David Rittenhouse* (Princeton, N.J., 1964).

servation of the transit of Venus. The product of scientific labor, the published papers on the transit, reveal Rittenhouse's activity in science as different in kind from his clockmaking or his manufacture of surveying instruments for sale. The flexibility of the American structure allowed this movement from an artisan measurement system to a scientific one.

Rittenhouse's fame as a scientist meant that his measurement skill was called on to perform practical tasks in behalf of the general public. Thus he became a surveyor of state boundaries, a builder of fortifications, a financial official of the new state of Pennsylvania, and finally the director of the Mint. From one point of view, this was a waste of scientific talent, but, from another, it was a recognition of the relevance of measurement talent to the problems facing the new Republic. If Rittenhouse showed himself to better advantage as a surveyor than as a financier or an administrator, it is an indication both that the transfer of measuring activities from one setting to another was not always efficient and that the American Republic was in desperate need of measurers. A scientific reputation was a clear signal of measuring ability, and the best way to move measuring ability from a scientific to a technological setting was by co-opting the person. The fulfillment of a scientific career often involved being redirected toward tasks relevant to social demands.

Americans were a people obsessed with measuring land; this interest permeated all levels of the population and drew the talents of many of the nation's ablest leaders. The grid of the Land Ordinance of 1785 has left a lasting mark on many parts of the United States, although the older system of metes and bounds has, in certain instances, adapted advantageously as well.[14] By the time of the Revolution, Americans were sophisticated measurers. The prominence of surveying in the education of Abraham Lincoln is an indication of the penetration into the population of the Rome-to-

---

14. Norman J. W. Thrower, *Original Survey and Land Subdivision: A Comparative Study of the Form and Effect of Contrasting Cadastral Surveys* (Chicago, 1966). For a preliminary analysis of the Rome-to-America grid, see A. Hunter Dupree, "The Pace of Measurement from Rome to America," *Smithsonian Journal of History,* III (1968), 19–40.

America measuring grid.[15] From the beginning, the United States was not underdeveloped in measurement activities.

The Rome-to-America grid, though modified both by the Renaissance and the Enlightenment, is one of several striking continuities with the Middle Ages which have given a special character to the American frontier.[16] Land measurement has been a pervasive activity throughout the history of the Republic. The surveyor-measurer has persisted in every community, as has the profession of civil engineer, which shared the same origin. Land measurement also required the services of astronomers to establish base latitude and longitude. Even the adjustment of the grid to the curvature of the earth required an element of sophistication akin to scientific technique rather than simple surveying.

FOUR SYSTEMS OF COORDINATES

Nineteenth-century measurement systems in America, however, were not confined to the rectangular grid method of land measurement. From the early Republic, a number of other systems of coordinates were available, and Americans knew how to use them. Spherical coordinates for the surface of the earth and the corresponding celestial sphere played an important role in nineteenth-century America. Navigation was essentially organized as a craft, with apprentice education and highly stable templates. Even Nathaniel Bowditch, in his early career, was essentially interested in improving the accuracy of a stable body of data. When he interested himself in Laplace he stepped from technological measurement to scientific measurement.[17]

The nineteenth-century exploration-survey expeditions both on

---

15. Two surveying books used by Lincoln were Abel Flint, *A System of Geometry and Trigonometry* (Hartford, Conn., 1808); and Robert Gibson, *A Treatise on Practical Surveying* (Dublin, 1810). I am indebted to Mrs. Christine D. Hathaway, Special Collections Librarian, Brown University, for this information.

16. White, "Legacy of Middle Ages," pp. 191–202.

17. Dirk J. Struik, *Yankee Science in the Making* (New York, 1962), pp. 230–32.

land and sea were essentially concerned with measuring by spherical coordinates. The expeditions manned by Topographical Engineers and naval expeditions such as that of Charles Wilkes were measuring in a tradition that owed much both to astronomy and to navigation. A West Point education, with its emphasis on French mathematics, had relevance in preparing engineers for these expeditions.

Two more sets of measuring coordinates completed the repertoire available in the survey-exploration tradition of the nineteenth century: biological classifications and geological dating. They are not conventionally included as mathematical coordinates, yet one can conceivably interpret them as such. Certainly they had almost as prominent a place on the surveying and exploring expeditions of the nineteenth century as did the rectangular and spherical grids.

The much-maligned Linnaean system for classification of plants and animals and its successors—the so-called natural systems—were mathematical albeit untouched by the tradition of analysis. The botanist of the early nineteenth century had much in common with a measurer of the early Middle Ages. He thought geometrically rather than algebraically, and he utilized words to summarize groups of geometrical shapes—morphologies—which could then be categorized in series of increasingly inclusive categories. Once placed by binomial nomenclature into species and genera, the groups of morphological characters, measured geometrically and symbolized by language, could be manipulated statistically. The quest of the natural system was for affinity—a relating of each group to all those which it most nearly resembled—suggesting a complicated, three-dimensional matrix. Despite the failure of the quest for a natural system, the naturalist of the nineteenth century could use classification to place the flora and fauna of North America in rough order. Since the known species by mid-century were around 100,000, the elaboration of the flora and fauna was a very large task. John Torrey, who embarked on a *Flora of North America* in the 1820s, would doubtless have been incredulous if told that such a work would still be incomplete in 1970. The naturalists of America in the nineteenth century were driven men, most of whom died with their programs unfinished because there was no way available to them to group their results except by simple addition.

The geological time scale was a combination of a linear scale for strata of rock and a biological classification applied to fossils. The geologists of the surveys faced a formidable task in elucidating the formations of a continent. While certain practical possibilities such as the presence of minerals seemed to beckon to the scientists, the basic problem of the geologist was to unravel geological history; the prospecting craft could do a better job of locating ore deposits.

Both the biological and the geological measurement systems thus fell within the domain of science, not of technology. Since scientific literature, by definition, followed no time schedule, part of measuring had to do with keeping the abstract memory bank tidy. The housekeeping rules had to determine what information should be allowed into the corpus of science, and measuring behavior consisted of constant decision-making. The rule of priority was the main device for preventing redundancy of information, and the rules of Linnaean nomenclature provided a handy program for priority decisions. The natural history measurement systems were thus self-stabilizing and were highly intolerant of the introduction of names for organisms already described. Theirs was an entirely different kind of discipline from that imposed on the farmer, who had to make a crop every year.

DARWIN AND THE FOUR SYSTEMS OF COORDINATES

The single greatest act of creative genius in the nineteenth century was performed by Charles Darwin. He did not think of a new idea, or even innovate in a particular branch of science. Rather he combined the measurement practices of the four scales mentioned here to apply to the total stock of living organisms on the earth. The surveyors' and navigators' activities combined with the biologists' concerns produced a world-wide measure of the geographical distribution of organisms. The dimension of the geological time scale completed Darwin's mathematical synthesis, which underlay and survived all the metaphors by which he and others described the system. Four systems of measurement, all arranged in ways which suppressed variation and tended to stability, came together to ex-

33

plain variation and to introduce one of the most striking changes of perspective in the history of science.

Americans in the relevant measuring groups were so busy with their local programs that they appeared backward when compared to Darwin. Yet his relation to Asa Gray demonstrates that Darwin was as exceptional on the European as on the American side of the water and that a prime requisite for an American's participation in scientific innovation on Darwin's scale was relief from commitment to his ongoing programs. Gray would have been glad to take up geographical distribution of plants in the early 1840s. It was only when Darwin told him to forget about the *Flora of North America* and to concentrate instead on some statistics in the imperfect *Manual* that Gray entered the Darwinian enterprise. For a brief time he showed marked ability at abstract thinking and was one of the few naturalists to apply Darwin's measuring synthesis to a particular biological situation (in his case, the relation of the flora of eastern Asia and eastern North America) and thereby provide—even before the publication of the *Origin of Species*—empirical confirmation for the transmutation of species.[18]

The aftermath of Darwin's synthesis, however stimulating to those disciplines in biology such as physiology and cytology, which had not hitherto been a part of the surveying-exploring complex, left the measuring groups of the earlier period unrevolutionized and still pursuing their old programs. In Europe, Bentham and Hooker embarked on a nonevolutionary *Genera Plantarum*. In America, Asa Gray spent the remainder of his life trying to complete the *Flora of North America*.

In geology the post-Darwinian program was more influenced by the new viewpoint, but as in the case of Clarence King the elaboration of the Western sequences could be done as well by a catastrophist as a uniformitarian evolutionist. John Wesley Powell showed perhaps more awareness of the four measuring systems than any other scientist in the trans-Mississippi West after the Civil

---

18. A. Hunter Dupree, *Asa Gray: 1810–1888* (Cambridge, Mass., 1959), pp. 233–63.

War.[19] He attacked the rectangular-grid concept as embodied in the 160-acre homestead and looked instead for political boundaries to coincide with drainage basins of rivers. He saw as a single system the relation of man, land, and weather in the Colorado basin. Yet the work in both geology and ethnology which constitutes his lasting monument was part of a traditional program of research.

THE LABORATORY AS A SETTING FOR MEASUREMENT

Sometime in the late nineteenth century the scientific exploration-survey behavior setting, with its supporting institutions—the museum, the herbarium, the astronomical observatory—went into an eclipse behind the new setting for science—the laboratory.[20] When the scientist was content to sample the systems in the environment and watch them work unmolested, he was playing the role which Roger Barker has labeled "transducer." In this role, the investigator tries above all *not* to disturb the system that he observes and transforms into data.[21] On the other hand, the scientist may adopt the role which Barker calls "operator." Here he sends messages into the environmental system and then transforms the output into data. He can select fewer phenomena for measurement and can repeat the operation at will. He must simplify that part of the system which he observes and control his experiment by excluding most of the phenomena which the other tradition seeks to classify. In other words, he must set up a laboratory.

The rise of the laboratory is one of the great unwritten chapters in the history of science; by the end of the nineteenth century it had become the dominant institutional home of the scientific com-

---

19. A. Hunter Dupree, *Science in the Federal Government: A History of Policies and Activities to 1940* (Cambridge, Mass., 1957), pp. 195–214. See also Wallace Stegner, *Beyond the Hundredth Meridian: John Wesley Powell and the Second Opening of the West* (Boston, 1954); and Thomas G. Manning, *Government in Science: The U.S. Geological Survey 1867–1894* (Lexington, Ky., 1967).

20. "Behavior setting" and other terms in this chapter have been adapted from Roger G. Barker, *Ecological Psychology: Concepts and Methods for Studying the Environment of Human Behavior* (Stanford, Calif., 1968).

21. *Ibid.*, pp. 140–41.

munity in America. It increased the scientist's control over his experiment and increased his ability to keep other professionals—military men and lawyers as surveyors, for example—completely away from his professional habitat. The rise of the university laboratory and of the ideal of pure research divorced from all applications to technology are associated with science's coming indoors into the laboratory. Even astronomical observatories in the hands of astrophysicists (George Ellery Hale, for example) became laboratories at the near end of the telescope.[22]

MEASUREMENT IN THE UNITED STATES AT THE END OF THE CENTURY

As one looks back on the panorama of American history in the nineteenth century, measurement systems stand out as a prominent part of American culture both in science and in technology. By 1900, they had gained not only organization and support, but also had developed technique and sophistication. They had offered a marked increase in the options available to the society for solving its problems. Many spokesmen for the scientific community insisted on their separateness from technology and from practical goals. Yet they believed that if scientific information and skill were made available to the ancient crafts, even more dramatic improvements in the efficiency of agriculture and industry would be forthcoming. So confident were some enthusiasts that they saw science as a religion and a rule of life completely superseding all other traditions from the past. A sober look at industry and agriculture in 1900 suggests that their belief in the final triumph of science was an article of faith, not an accomplished fact.

In conclusion, we can venture a purely tentative general rule. *Measurement groups tend to seek stability and suppress innovation.* This proposition is accepted when applied to technological groups, though even here the myth of progress dies hard.[23] Yet the scientific community, in most situations, seeks stability also. The rule of

---

22. See Helen Wright, *Explorer of the Universe: A Biography of George Ellery Hale* (New York, 1966).

23. Cf. Donald A. Schon, *Technology and Change* (New York, 1967).

priority means that only new items are added to the cultural memory bank, but they are even more welcome if they do not jostle the pre-existing array of information. Certainly the groups which must always allocate scarce resources for information-generating activity will consistently be tempted to choose to invest in data production which will improve the already existing matrix.[24]

If measuring groups tend to foster stability, whence comes movement in a century whose name is change? Accounting for change is always a problem in institutional and structural-functional analyses.[25] A brief answer may be found in a corollary to the rule of stability. Changes in behavior of both scientists and technologists come about by invasion from outside the group. Changes in the state of the information system, however, can be produced by stable behavior patterns. Science becomes exponentially capable of change within its own domain. In earlier times innovation in the craft groups arose by invasion of upstarts from the general population who were called inventors. As the nineteenth century wore on, however, the invasion of the technological groups increasingly came from the scientific community. Meanwhile the scientists themselves enthusiastically invaded other scientists' stabler bodies of knowledge. Out of the sometimes acrimonious collisions came the innovations which ultimately changed the pattern of American culture.

---

24. Cf. Thomas S. Kuhn, *The Structure of Scientific Revolutions* (Chicago, 1962). Where I differ is that I try not to describe that activity which Kuhn calls normal science in pejorative terms.

25. This may be true of the structural-functional sociologists, for example, Talcott Parsons, Robert Merton, and Bernard Barber, with whom I feel more in common than I do with any historian of science.

American Indifference to Basic Research: A Reappraisal

American indifference to basic research, decried both in academic circles and among the educated public, is an old theme in the literature of national self-appraisal dating back at least to de Tocqueville. We Americans abjectly trot it out when bested by Europeans. When Soviet Sputniks orbited the earth, American indifference to basic research was blamed for this technological defeat. When we landed on the moon, quite properly no praise was given to basic research. In recent years, American science—the product of this supposed indifference—was castigated for neglecting the problems of our society. Almost forgotten are the enormous successes achieved by a research and development community highly responsive to national needs.[1]

The American research and development community is the envy of the world and the object of fascinated, avid study by science policy experts in international and foreign national bodies. Obviously, American practical achievements first hit the eye. Practical results are not necessarily incompatible with indifference to basic research—the Japanese have done very well economically with relatively little basic research. But ironically enough, of all Western nations the indifferent Americans devote a greater percentage of their gross national product to research and development in general

---

1. A recent and quite superior example of this concern with science and technology in the United States is the report of the Organization for Economic Cooperation and Development, *Reviews of National Science Policy: United States* (Paris, 1968).

and probably to basic research in particular.[2] What does this mean historically? The purpose of this question is neither to deny nor to confirm American indifference. I do not know if Americans were indifferent to basic research, and neither does anyone else. This widely discussed topic has become a sterile historiographic concept lacking adequate evidence for either refutation or confirmation.

Beliefs affect perceptions of reality. As these perceptions were applied to the evidences of the past, scientists and historians concerned with the sciences in America inevitably produced writings reflecting their attitudes toward the indifference theme. Two related tasks are proposed here: first, a survey of the characteristic ideas or criteria in the indifference literature to suggest some of the underlying connections to the real world that these writings purport to describe; second, suggestions of how we might find out more about the relationships of Americans and research in order to predict what these relationships might be.

The extensive literature on American attitudes to basic research goes back to the early years of the Republic and is absorbing reading, even to educated laymen interested primarily in our current dilemmas.[3] Besides de Tocqueville, there is a host of foreign commentators who failed to find their image of a cultural center in the United States. Their laments and similar American ones are clearly related to the assumption that Americans were not interested in literature, music, and the arts. Someone might profitably compare the artistic and the scientific laments. Although overtaken by events, these assumptions still persist on both sides of the Atlantic. Around 1876, the centennial brought forth a number of articles on America's inferior position in the scientific world. The end of the century was another time of national stocktaking in which our scientific deficiencies were publicly decried. In both 1876 and 1900, the critics were

---

2. The United States was apparently the first nation to expend over 3 per cent of GNP on research and development. See *ibid.*, p. 32. (At the time, Soviet data were not available in a readily comparable form.) Clearly, the 1969 estimated support of basic research, $3,730,000,000, is the largest in absolute terms, and perhaps relatively as well. See National Science Foundation, *National Patterns of R & D Resources* . . . (Washington, D.C., 1969).

3. There is no complete bibliography of such works. The best published starting point is I. Bernard Cohen, *Science and Society in the First Century of the Republic* (Columbus, Ohio, 1961), with its useful bibliography.

usually scientists. American scientists have always decried American callousness toward the search for truth. Foreign experts have sided with the scientists.

Historians have accepted the scientists' criticism as fact. Unconcerned about the sciences, most have probably been convinced that the country was not really interested in scholarship and art. When historians of the sciences appeared in this generation, they found the writings of Richard H. Shryock and I. Bernard Cohen affirming the scientists' complaints.[4]

Richard H. Shryock started out in American political history but soon became interested in medicine, and public health in particular, as well as in more general questions of American social and intellectual history. His work in medical history has a very strong concern with social context. Aspects of medical science and practice in the Shryock canon are often examined in terms of social factors which affect application. Shryock has investigated how scientific advances impinge on medicine and how nonintellectual forces determine the acceptance of scientific developments. His work has had considerable influence not only because of its quality but because he has always tried to relate his findings to broader questions in American history and in the history of science.[5]

In 1948, Shryock published an article entitled "American Indifference to Basic Research During the Nineteenth Century."[6] It is the classic article; when Richard Hofstadter wrote on American anti-intellectualism, Shryock was cited to make the point about basic research.[7] The structure of Shryock's piece is simple. After asserting the fact of general indifference, Shryock tests and rejects various alternative explanations of the phenomenon: availability of Euro-

---

4. For reactions of two leading postwar scholars, see A. Hunter Dupree, "The History of American Science—A Field Finds Itself," *American Historical Review*, LXXI (1966), 863–74; Edward Lurie, "An Interpretation of Science in the Nineteenth Century: A Study in History and Historiography," *Journal of World History*, VIII (1965), 681–706.

5. See Richard Shryock, *Medicine in America: Historical Essays* (Baltimore, Md., 1966). A bibliography of Shryock's writings is in the *Journal of the History of Medicine and Allied Sciences*, XXIII (January, 1968), 8–15.

6. Originally in *Archives internationales d'histoire des sciences*, XXVIII (1948), 50–65.

7. Richard Hofstadter, *Anti-Intellectualism in American Life* (New York, 1963), p. 26.

pean science; preoccupation with conquering a continent; and clerical control of colleges and universities. Accepting a fourth thesis—extreme emphasis on utility—Shryock asserts that preoccupation with utility was not evident in the colonial period and became increasingly apparent only as the nation industrialized. The indifference ultimately stemmed from the industrial and mercantile leadership. When Americans recognized the implications of science for technology in about 1900, we have, in Shryock's words, "the rather sudden emergence of basic science in the United States."

Shryock's explanation does not seem tenable. Take the two most spectacular philanthropists of the early twentieth century, Carnegie and Rockefeller. The former did support technical education but backed pure research in the Carnegie Institution of Washington. The Institution has rarely deviated from pure research. There is ample evidence that this is what Carnegie and his trustees wanted. The principal Rockefeller aid to research was not in physics, chemistry, and geology, which might ultimately have redounded to the benefit of petroleum industry holdings, but in medical research and public health. Even when the various Rockefeller endowments broadened their scopes, one could argue that concern for cultural life was at least as strong a motivation as any desire for ultimate applied benefits. The point here is that there is, indeed, very little evidence which indicates that Carnegie and Rockefeller realized the implications of science for technology and pumped money into science. The sources of support for science prior to the massive infusion of federal funds starting in World War II are so varied as to cast doubts on any explanation stressing a single factor. The industrial and mercantile elite was as likely or more likely to put funds into stadiums and art museums as into laboratories.

I. Bernard Cohen was originally a mathematician. A student of George Sarton, who was the pioneer historian of science in America, he succeeded his teacher as editor of *Isis,* the most eminent journal in the field. Cohen has published extensively on many topics and has trained a significant number of the historians of science in America. Although discussed here in the Americanist context, Cohen is essentially a historian of astronomy and physics with emphasis on the seventeenth and eighteenth centuries. He is not primarily concerned with social context, that is, the external history of science. On the contrary, Cohen is a leading practitioner and proponent of the internalist history of science—the study of the development of scientific concepts and of the scientific knowledge ordered by these concepts. Social and other contexts are decidedly secondary in intellectual interest to these internal developments.[8]

The internalist position is—to oversimplify—a reaction against Marxian determinism as proposed in the 1930s and the somewhat naïve sociological approach of Robert Merton's "Science, Technology, and Society in Seventeenth Century England," [9] in favor of the intellectual history of the context of science, particularly its conceptual aspects. Most of the dispute in the 1930s centered around the origins and nature of the scientific revolution of the seventeenth century.[10] American events played no role in this dispute. With the general victory of the intellectualist historiographical approach, American events were largely neglected by internalists since the indifference literature already affirmed the relative insignificance of American scientific achievements from a conceptual standpoint. What historical work did occur largely stemmed from men like Shryock, the elder Schlesinger, and Merle Curti, who were, broadly speaking, American social and intellectual historians.

The differences between the two schools are describable in terms of three foci of concern: the contents of science, the scientists themselves, and the society at large. The orthodox internalist historian of science is primarily concerned with the contents of science and only with those parts of the scientists' lives that influenced their professional work (i.e., religion, general philosophic beliefs, institutional affiliations). The conventional historian is concerned with aspects of the society at large. Shryock and others studying the sciences in America are in the uncomfortable position of being primarily concerned with the scientists as such, and, moreover, wanting to understand the scientists in terms of both the scientific content and the

---

8. See *International Encyclopedia of the Social Sciences,* s.v. "Science, history of."

9. *Osiris,* IV (1938), 414–565.

10. A good introduction to the controversy is George Basalla, *The Rise of Modern Science: Internal or External Factors?* (Lexington, Mass., 1968).

social environment.[11] Cohen, nevertheless, has produced an appreciable body of writings concerned with the sciences in America. Unlike Shryock, he does not venture far into social or cultural history. Like Shryock, he stresses the debilitating effects of pressures for practical results.[12]

In addition to their stress on the pernicious effects of practical pressures, the two men agree on other aspects of the indifference theme. Both pay considerable attention to the absence of great men and great contributions to major advances in the sciences. As is typical in this genre, they usually cite Benjamin Franklin, and then note the exceptional cases of Joseph Henry and J. Willard Gibbs. Both men also stress the small number of American basic scientists in the past.

Shryock and Cohen both emphasize colonial science. Both, viewing the colonial period almost as a golden age, attribute its scientific fecundity to an aristocratic, or at least patrician, sympathy for basic science. From this position is derived a view of the history of science in America as being like a Bactrian camel, a nineteenth-century trough between colonial and twentieth-century humps. Shryock and Cohen have both written of the colonial period and are understandably enthusiastic about it. However, Franklin, John Winthrop, John Logan, David Rittenhouse, and Cadwallader Colden hardly consti-

---

11. For an Americanist-internalist confrontation, see Shryock's comments in "Conference on the History, Philosophy, and Sociology of Science," *Proceedings of the American Philosophical Society,* XCIX (1955), 327–54; and Herbert Dingle, "History of Science and the Sociology of Science," *Scientific Monthly,* LXXXII (1956), 107–11.

12. Cohen's work includes: *Benjamin Franklin's Experiments . . .* (Cambridge, Mass., 1941); "How Practical Was Benjamin Franklin's Science?" *Pennsylvania Magazine of History and Biography,* LXIX (1945), 284–93; *Franklin and Newton . . .* (Philadelphia, Pa., 1956); *Some Early Tools of American Science: An Account of the Early Scientific Instruments and Mineralogical and Biological Collections in Harvard University* (Cambridge, Mass., 1950); "Some Reflections on the State of Science in America during the Nineteenth Century," *Proceedings, National Academy of Sciences,* XLV (1959), 666–77; "La Vie scientifique aux Etats-Unis au XIXe siècle," in R. Taton, *Histoire générale des sciences,* III, 633–44; "Science in America: The Nineteenth Century," in *Paths of American Thought,* ed., A. M. Schlesinger, Jr., and Morton White (Boston, 1963), pp. 167–89; *Science, Servant of Man* (Boston, 1948); "American Physicists at War: From the Revolution to the World Wars," *American Journal of Physics,* XIII (1942), 233–35, 333–46; "Science and the Revolution," *Technology Review,* XLVII (1945), 6; "Science and the Civil War," *Technology Review,* XLVIII (1946), 5.

tuted a scientific community. With all due allowance for Franklin, a great man in many respects, the overall colonial scientific output was not very large or significant.[13]

Enthusiasm has often overpowered critical faculties. Consider, for example, the question of the number of Americans who were made fellows of the Royal Society. That there were many more in the colonial period than later is offered as proof of the existence of a colonial hump. Forgotten is the fact that colonials were British subjects; after 1783, Americans had to become foreign fellows, a severely restricted honorific category. Further, the standard source of information on colonial fellows lists West Indians, a Canadian, and even Elihu Yale, who was born in Boston but lived his adult life in the Eastern Hemisphere. Eighteen bona fide American colonial fellows from 1660 to 1788 seem a very small hump, indeed, considering that many of them made few scientific contributions at all.[14]

To return to the more general characteristics of the indifference literature, periodization is a problem to many authors. What date does one use to mark the point when indifferent Americans became favorably disposed to basic research? Is it Johns Hopkins University's effective launching of graduate education in the United States in 1876? 1900 is a possible candidate, probably because it is easy to remember. World War I is another alternative, because American strengths became more apparent in contrast to Europe's distress. The arrival of the refugees fleeing Hitler is a conceivable turning point. The refugees, after all, were notable as mathematicians and physical theorists, specialists in areas in which Americans were weak in comparison to Germany. If drama and social significance are the criteria,

---

13. In Philadelphia, especially, and in Boston there were signs of emerging small groups of the scientifically minded toward the end of the colonial period. This could hardly be compared in scale and quality with events in the next century; for example, compare the Harvard and Yale faculties in 1760 and in 1860. A recent dissertation estimated that Americans in the years 1800–60 produced approximately 10,000 scientific papers. Donald Beaver, "The American Scientific Community, 1800–1860: A Statistical-Historical Study" (Ph.D. diss., Yale University, 1966), p. 102.

14. R. P. Stearns, "Colonial Fellows of the Royal Society of London, 1661–1788," Osiris, VIII (1948), 73–121. Stearns's purpose, of course, is not to list merely those from the thirteen original colonies but from all of British North America (including the West Indies), for he is concerned with the total intellectual traffic across the Atlantic.

1945 and the first atom bomb at Alamogordo could signal the end of indifference. Obviously, the date selected is arbitrary. Not so obvious is the frequent assumption of a historical discontinuity, that two distinct periods exist and that little or nothing prior to the turning point requires consideration in explanation of the present situation. Evidences of this continuity in the history of science in America are quite plain; a good number of native scientific institutions antedate the Civil War, and our university pattern was in existence before 1900.

A more important characteristic of the indifference literature is the loose use of words like *basic* and *pure* as modifiers of *science*. Many authors, unaware of semantic difficulties, use the two interchangeably. *Basic* refers to intrinsic merit, usually scientific activities involved in formulating and verifying hypotheses and general theories. *Pure,* in contrast, refers to a psychological motivation unsullied by concerns other than the growth of scientific knowledge. Stress on the pernicious effect of the American ideal of practicality on science leads authors to the assumption that purity yields the basic and that the existence of the basic is proof of purity. In practice, the situation is not quite so simple. How does one determine purity of motive? How does one classify Irving Langmuir, a great theoretical physical chemist, who worked at the General Electric Research Laboratory and even took out patents? What do we do with a university physicist who consults with the Atomic Energy Commission on the development of nuclear power? And what do we do with the very many pure-minded scientists on countless campuses whose contributions to the development of basic theory will never amount to even a footnote?

Related to this semantic confusion is the assumption that we can always or usually clearly differentiate between some activities designated as science and others designated as technology, invention, or development. No writer in the indifference literature explicitly and seriously raises the point that research situations are quite often mixed. In addition to the problem of differentiating purity of motive and intrinsic merit, there is the problem of differentiating science (theory and its underlying data) from technology (applications meeting human needs). A mixture of scientific and technological concerns typified a significant proportion of research activities in the past; quite clearly, such mixtures exist today.[15] The existence of "pure" or "basic" scientists, and of practical men concerned solely with solving immediate problems, is not being denied. What is being stressed is that the significant presence of mixed situations renders analysis in terms of basic and applied science increasingly sterile. For convenience, we establish arbitrary distinctions, but the results often do violence to the real world under discussion. Writers convinced of the truth of American indifference to basic research will enthusiastically seize on every kernel of practicality to downgrade scientific contributions in the American past. Historians of the opposite persuasion will exalt every nugget of theoretical concern as proof of the scientific glory of the American past.[16]

As George Daniels points out, many indifference-theme authors, discussing motivations that lead to applications, make no distinction at all between the scientists and society at large.[17] Society may sponsor research for presumedly practical reasons, while the researchers themselves are often primarily interested in expanding knowledge. An example is the case of a great paleontologist-geologist of the last century, James Hall.[18] The New York State Survey, which he headed, undoubtedly originated in large measure in the concerns of many for more accurate information about exploitable natural resources. Yet, year after year, Hall turned out a series of abstruse volumes of little apparent applicability. The state legislature grumbled; Hall vented his ire at his critics; but the money continued to come in support of the research. Hall did not convert the Philistines

---

15. See Nathan Reingold, "Alexander Dallas Bache: Science and Technology in the American Idiom," *Technology and Culture*, XI (April, 1970), 163–77; *idem*, "Cleveland Abbe at Pulkowa: Theory and Practice in the Nineteenth Century Physical Sciences," *Archives internationales d'histoire des sciences*, XVII (1964), 133–47; Charles Davies, *The Logic and Utility of Mathematics . . .* (New York, 1850), pp. 3–4, a nineteenth-century statement of this view.

16. In recent years the problems arising from the terms *pure* and *basic*, not to mention *applied* and *development*, have been met somewhat by the device of postulating "mission-related basic research." Part of the problem arises from different ways of looking at research because of different needs (i.e., where funds originate, where they are spent, who performs the work).

17. George H. Daniels, *American Science in the Age of Jackson* (New York, 1968), p. 21.

18. See John M. Clarke, *James Hall of Albany* (New York, 1922).

46

to the true faith of science. He was just tough enough to survive until his work was taken for granted as part of the overhead of state government.

In complaining about the paucity of great men and great contributions to major scientific developments in the United States, the authors of the indifference literature pay particular attention to the physical sciences, especially astronomy and physics. Even authors interested in the biological sciences, such as Shryock, reflect this awe for the Comtean hierarchy. When Shryock and others consider biology, the stress is on the lack of American contributions to the cell theory, to the germ theory of disease, and to physiology, all areas presumably characterized by a respectable degree of experimentation, quantification, and theoretical structure. Anti-intellectualism, they assume, leads to a mistrust of abstract or theoretical work. As a consequence, the argument goes, Americans tended to go in for taxonomic fields—natural history and geology. Although neither experimental nor quantitative, a taxonomy is, after all, a theoretical statement of relationships. Nor is the work involved in developing a taxonomic scheme necessarily Baconian data-grubbing.

The Baconian position is ahistorical and distorts the nature of the past research communities. One picks out high points in contemporary science, looks backward in time to establish a "main line" of development, and largely blots out all research not in or near the "main line"—physical theory and experimental biology. Geology, a very important field in nineteenth-century Europe and America, is ignored, except in relation to evolution. A conventional concern of historians of science is to overturn the accepted description of sections of the "main line" by pointing out omissions of additional intellectual factors. Because the indifference theme has a social context, it apparently never occurs to most historians of science to question the oversimplification of the "main line" thesis in studying science in America.

Take the example of mathematics. This field is one of the presumed weaknesses of the national scientific community, as one would expect of people who are pictured as anti-intellectual, not given to abstract, theoretical work. As far back as 1940, L. C. Karpinski published a magnificent bibliography of mathematical works

47

printed in America to 1850. He lists 1,092 appearing in a total of 2,998 editions. (These include many reprints and translations of leading European works.) Karpinski concludes:

No reasonable person should expect that in the period before 1850 America would have produced mathematicians to rank with Newton and the foremost European scientists. It was too much occupied with settlement, exploration, and expansion. The wonder is that at this time the population could furnish a body of readers of so much mathematical work that was really valuable. *Upon the work of these years America built,* and comparatively rapid progress after 1850 is due in no small measure to the broad and solid foundation laid in the earlier period.[19]

The wonder is that historians have not reconsidered glib assumptions about Americans and exact sciences. While Americans prior to 1850 or just afterwards did not spawn schools of great mathematicians, it is true that the evidence requires us to rethink our positions on many aspects of the history of science and technology in the United States. Would we have a different historiography if our predecessors had started from the books of Karpinski and the thought of such men as Nathaniel Bowditch; Bowditch's student, the mathematician-astronomer Benjamin Peirce; and Peirce's students, Charles Sanders Peirce and the astronomer Simon Newcomb? Suppose they had been impressed by William Ferrel's work on the mechanics of the earth's atmosphere?

Most indifference-theme authors implicitly compare the United States with all of Europe, or at least with all of Western Europe. A comparison of the two as they were in 1800 is invalid. America and Europe were probably not comparable as late as 1900. At issue are not only relatively simple matters of population and gross national product, but intangibles such as the nature of educational systems, the quality of scientific institutions, the efficiency of scientific communication, and, indeed, the nature of the entire relationship of the

---

19. L. C. Karpinski, *Bibliography of Mathematical Works Printed in America through 1850* (Ann Arbor, Mich., 1940), p. 17.

scientific community and the society at large. The implicit comparison with all of Europe is not based on any findings of statistical equivalency; it arises from a pattern of unreasonable expectations. For example, in Denmark, the succession Brahe, Römer, Oersted, Bohr is roughly one significant physical scientist per century,[20] the same average as in the United States of the eighteenth and nineteenth centuries with Franklin and Gibbs. I am not aware of any breast-beating or finger-pointing literature about science in Denmark. Neither the Danes nor anyone else expects Denmark to dominate or to match the other nations in scientific fecundity. As a matter of fact, some smaller European nations contribute more to science relative to their population than larger powers. Obviously, a different pattern of expectations is applied to the United States.

The indifference charges often have strongly nationalistic undercurrents. Not only should the United States participate significantly in the great human achievement known as science, but it should lead. One can understand why scientists would want to see an expansion of science in America and even seek the pride of first place. It is unclear how historians of science who stress the autonomous development of the sciences largely, if not wholly, determined by the internal logic of the sciences, can point to numbers and wealth as reasons why America should have or might have contributed more to science. Nor is it clear what is so strange or horrendous about Americans not being among the founders of the cell theory or any other major scientific development. There are sometimes rather prosaic explanations of the improbability of such American contributions. Except for nationalistic pride, is there any reason why Americans or any other nationality *must be* among the founders of any major scientific concept? As the number of original theorists is necessarily small, an honest historiography will disappoint many nationalisms.

Perhaps the strangest characteristic of the professional historians'

---

20. T. Brahe (1546–1601), the astronomer; O. Römer (1644–1710), the astronomer; H. C. Oersted (1777–1851), the natural philosopher; N. Bohr (1885–1962), the physicist.

reactions to scientists' complaints of public indifference is the former's lack of critical scrutiny. Much modern historiography of the sciences originates in attempts to complete or to correct the traditional accounts in scientific literature. Many historians of science now consider these accounts spurious myths that serve to inculcate orthodoxies about the nature of science. The very men who will insist that students of the history of physics must scour the printed sources and rummage through bundles of old manuscripts blandly accept scientists' assertions on the indifference theme. Are the statements factually accurate? Were the authors self-seeking or otherwise motivated? Because past historians were predisposed to agree with the scientists' laments, they often misunderstood the context from which the writings originated and missed key developments. For example, from the studies produced—amidst charges of public apathy—near the turn of the century, no one could predict the development of genetics. Yet the quality of much American biological groundwork in 1900 is quite evident in historical perspective.

As to complaints about the lack of support and appreciation, are they confined exclusively to the United States? When haven't scientists complained? Observers from the "other culture" sometimes feel that the scientists are perpetually dissatisfied out of sheer greed and the drive for power. A more reasonable explanation is that scientists have tended to complain because of the growing pains of their disciplines. We can speculate that, since the time of Newton, the sciences have grown rapidly as a consequence of a series of major intellectual advances. Men attracted to the sciences were caught up in the sheer excitement of the advances which, in turn, attracted more men to science. At any given moment, support for the expanding sciences probably tended to lag behind needs. Strongly committed scientists were very conscious of the gap between needs and actual support. Not surprisingly, they complained. We can even speculate that complaints are a sign of a healthy, lusty scientific community. Their absence may denote dangerously smug self-satisfaction, stagnation, or actual decline. In retrospect, we can see that when the French scientists stopped complaining early in the last century and the Germans stopped complaining toward the end of the century, both were entering periods of relative decline. The British, to their credit, have complained fairly steadily since at least the start of the last century.[21]

If we want to trace accurately the development of the sciences in America and what relations, if any, existed between the sciences and other aspects of American life, we have to avoid the easy generalizations engendered by the indifference literature. Perhaps the best way to start is by making two observations.

The first is that, so far as we can judge by available evidence, in the modern period applied research has always outbulked pure or basic research. This is true today in every major scientific community; it was apparently true from the time of Newton to the immediate past and in every country of any scientific consequence.[22] Providing health, foods, and goods understandably occupies much of the energy of mankind, including researchers. The indifference literature, animated by an almost aristocratic disdain, ascribes American applied work to sordid motives. Such motives do exist; however, many of the problems of improving health, food production, and technology are quite challenging to inquiring minds. These challenges may arise from enlightened concerns, as well as from sordid motives. In any case, realism indicates that very few men are endowed with the ability to perform truly basic research.

The second observation is inspired by a very good popular history of astronomy published in 1885. Citing the rise of astronomical research in the United States, the author notes the existence in 1882

---

21. See D. S. L. Cardwell, *The Organisation of Science in England: A Retrospect* (London, 1957). Interestingly, British complaints to this day are fundamentally different from American complaints. Where Americans have complained about the lack of support for the advancement of knowledge, the British have usually complained about the lack of support for the proper application of science.

22. The volume of publications is a good indicator. An examination of distribution of funds would also be helpful, but comparable historical data are lacking. For the current publication situation, see Charles P. Bourne, "The World's Technical Journal Literature: An Estimate of Volume, Origin, Language, Field, Indexing, and Abstracting," *American Documentation*, XIII (1962), 159–68. See also David A. Kronick, *A History of Scientific and Technical Periodicals* . . . (New York, 1962), *passim*; H. C. Bolton, *A Catalogue of Scientific and Technical Periodicals, 1665–1895* . . . (Washington, D.C., 1897); and W. A. Smith et al., *World List of Scientific Periodicals Published in the Years 1900–1950* (New York, 1952).

of 143 observatories.[23] These ranged from world-famous institutions to rather minor establishments. Some practical applications of astronomy did exist in 1882—time determinations and observations that related to surveying. These hardly justified 143 observatories. Until the space program, modern astronomy was very nonutilitarian. The large number of observatories certainly does not indicate an absolute indifference shattered only by a freakish Gibbs or Henry.

The underlying problem obscured by the indifference literature can be stated as follows: Given the existence of at least some basic research in the United States and the general preponderance of applied concerns over theoretical concerns, in what qualitative and quantitative respects did the American research experience differ from the European? If the differences are slight, we can characterize the American experience as a variant of the European situation; if the differences are substantial, two further questions arise: What are the origins of these differences? Are we faced merely with a gross variant or a new, unique pattern?

Of the qualitative aspects, three deserve mention: the great man issue, the nature of scientific work performed, and the distribution pattern of this work. Great men, great discoveries, and great traditions do exist and deserve careful investigation, but a historiography which is overwhelmingly based on these obscures the reasons why men, discoveries, and traditions are great and, most important, the processes involved in their genesis. Centering on greats guarantees a present-minded historiography making neat longitudinal sections backwards in time—the "main line" method. What we really want are successions of cross-sections in time, each showing the interrelations of various activities at a given moment. The sequence of cross-sections will enable us to formulate concepts explaining changes in time. Too many chance factors are involved in the appearance or nonappearance of great men for us to derive generalizations with any confidence. Who in 1900 could have predicted a Copenhagen school of physics in the twentieth century?

More promising is the study of the nature of the scientific work

---

23. Agnes Clerke, *History of Astronomy during the Nineteenth Century* (London, 1885), p. 8.

done in America. We have barely started on this line of investigation. The internalist historians of science possess skills that are essential, provided that they are supplemented by a critical concern for the total environment in which research occurs. Shryock's orientation, not Cohen's, is what we need.

Rather than judge scientific work in terms of nearness to a "main line," which has been determined in retrospect, it is necessary to understand clearly what occurred in a particular scientific investigation; what the origins of the investigation were; how the work fit into its contemporary scientific scene; and what, if any, its consequences were. Of these, the most important are the relationships to contemporary scientific work. Lonely precursors of later trends and geniuses towering across the ages are all too rare, and their appearance is based on too many chance hazards for fruitful insights into the norm of science at any time. The position of scientists and their work in terms of the standards of their contemporaries is what counts in judging the quality of the past scientific community. If this favorable judgment by peers is maintained and enhanced in time, we have a clear sign of the intrinsic strength of the scientific community and the presence of favorable conditions in the society for research. Changes in judgment are signs of change in scientific values.

Beyond the question of quality is the matter of the pattern of distribution of research effort. The indifference literature clearly postulates the existence of an excessive concern for descriptive, nonquantitative, and nonexperimental scientific endeavors in the American past. This type of research was not uniquely "American."[24] Some recent historical work clearly questions the overwhelming dominance of such scientific research in the American past.[25] Americans, after all, read European scientific journals and studied European monographs and textbooks, and a significant number in the last century went to European universities. More interesting is how

---

24. See Royal Society of London, *Catalogue of Scientific Papers,* 1st ser., 6 vols. (Cambridge, 1863–72).

25. Daniels, *Age of Jackson;* Beaver, "The Scientific Community"; P. A. Richmond, "A Selected Bibliography of American Fundamental Scientific Research during the Nineteenth Century" (M.A. thesis, Western Reserve University, 1956).

some fields rose earlier than others in America; how particular fields became ossified; and how others became favored recipients of society's largesse. The indifference literature suffers from being couched explicitly or implicitly in quantitative terms while lacking numerical foundations. "Less" and "more," "increase" and "decline," and "compared to" abound in the genre. Proving or disproving the assumption of indifference necessitates quantitative evidence.

If a colonial status for early American science is assumed, and if a subsequent sudden emergence of a metropolitan scientific pattern is further discerned, there really is nothing to analyze quantitatively. Colonial status in science can be defined not only as being small and dependent on metropolitan scientific communities for training, intellectual approbation, and basic conceptual tools, but also as partially replicating the metropolitan research and institutional patterns.[26] Being colonial in this sense does not mean doing particular and inferior kinds of science; such activities are also found in metropolitan scientific communities. A colonial scientific community is distinguishable from a small but mature nation's science by the fairly complete range of research and supporting institutions of the latter. Although participating in an international endeavor, the small but mature national scientific community will not have an almost complete intellectual dependence on others.

The colonial period in American science was more or less over by 1825. (This date and the following ones are not meant to imply abrupt, complete transitions, only approximations.) From that date to about 1875, science in America entered a crucial middle period in which the foundations of our national scientific community were established. By the terminal date, America was clearly a mature but small scientific nation with many of its basic institutions and at-

---

26. What follows consciously differs from the explanation of scientific colonialism given in George Basalla, "The Spread of Western Science," *Science*, n.s. CLVI (1967), 611–22. Specifically, the relevance of Basalla's analysis is doubtful in two respects. First, the stress on a different kind of science based on geographical newness is not basic. Similar activities occurred in Europe. Second, his analysis, with its strong emphasis on non-Western countries, does not apply to the United States. The settlers were of the West; science was part of the cultural baggage brought across the Atlantic. The concept of provincialism is preferable. The early colonists were in the same relationship to European centers of science as were Europeans in rural areas and the smaller towns and cities.

tributes in existence or in embryo.[27] Obviously, not all fields were equally strong. From 1876 to 1920 the scientific community developed further into a major world intellectual body. Opinions will naturally differ, but 1900 is a fair approximation of the date the United States became the scientific peer of any nation in Europe. Please note, *peer* does not mean *leader*. Sometime between 1920 and 1940, that role passed to this side of the Atlantic. The situation varied from discipline to discipline.

If these views are correct, we will not find a quantitative curve that shows a pre-nineteenth-century rise, followed by a relative decline and then an upsurge in this century. More likely is an unspectacular curve going from small (1825) to medium (1875) to large (1920) and then to very large (1960).

We need historical statistics for the number of American scientists and engineers, the number of articles and other written contributions, and the amount of financial support given to research. These should be further broken down by kind of scientific and technological research and by institutional location. But even these series are insufficient, giving rise to a fundamental problem, the difficulties of comparison. How can one prove, rather than merely assert, that Americans are more or less indifferent to basic research if there is no European yardstick? Only the preoccupation with greats and the emotional belief in anti-intellectualism as a basic feature of American life can explain this curious oversight in the writings of so many talented men. We need qualitative and quantitative data for *both* Europe and America to know if research patterns differ and in what degrees.

It is likely that future historical studies will show that the American research pattern in the last century was not overwhelmingly different from the as-yet-undetermined European norm. Recent his-

---

27. For the early period, see John C. Greene, "American Science Comes of Age, 1780–1820," *Journal of American History,* LV (1968), 22–41; *idem,* "The Development of Mineralogy in Philadelphia, 1780–1820," *Proceedings of the American Philosophical Society,* CXIII (1969), 283–95; and Leonard Wilson, "The Emergence of Geology as a Science in the United States," *Journal of World History,* X (1967), 416–37. For the middle period, see George H. Daniels, "The Process of Professionalization in American Science: The Emergent Period, 1820–1860," *Isis,* LVIII (1967), 151–66.

torical work in this area points to the discovery of much more basic or pure research than was predictable from the traditional accounts, in addition to the previously mentioned recognition of more scientific work of an abstract, theoretical kind (i.e., non-Baconian) than that indicated in the indifference literature. Deviations from the European norms are, in all probability, minor, and result from simple time lags or temporary local conditions. The American pattern is not unique, but a variant from the European norm.

A proviso is that the American data will differ most from the European data at the start of the nineteenth century, when the nation was still dependent on Europe for training, intellectual norms, and concepts. By 1860, the qualitative pattern and the gross quantitative data in the United States might be between those of the kingdoms of Prussia and Portugal. One can wave the flag slightly by predicting results somewhat closer to Prussia than Portugal. By the end of the century, the United States' commitment of a proportion of its gross national product to research and development was probably roughly equal to Germany's—perhaps even slightly larger. By 1920, the American commitment certainly justified comparisons with a war-ravaged Europe. Indirect evidence supports these assertions. The basic American university pattern was in flourishing existence by 1900. One can cite the favorable comments of foreign scientific visitors. A rather impressive number of new men and new institutions became prominent in the years 1890–1910.[28]

---

28. See the *Yearbooks of the Carnegie Institution of Washington, 1902–1916;* G. W. Corner, *A History of the Rockefeller Institute, 1901–53* (New York, 1964). For American higher education, see Lawrence R. Veysey, *The Emergence of the American University* (Chicago, 1965); Joseph Ben-David, "The Universities and the Growth of Science in Germany and the United States," *Minerva,* VII (1968–69), 1–35; S. M. Guralnick, "Science and the American College: 1828–1860" (Ph.D. diss., University of Pennsylvania, 1969). Among the favorable comments of foreign visitors are W. G. Waldeyer, "Relations between the United States and Germany, Especially in the Field of Science," *Annual Report, 1905,* Smithsonian Institution, pp. 533–47 (from the original in the *Sitzungsberichte des Königlichen Preussichen Academie der Wissenschaften,* Jan. 26, 1905); Maurice Caullery, *Les Universités et la vie scientifique aux Etats-Unis* . . . (Paris, 1917). For the 1920s, see League of Nations, Committee on International Cooperation, *Enquiry into Conditions of Intellectual Work, Second Series: Intellectual Life in the Various Countries: The United States of America* (Geneva, [1924]), Brochure No. 41. See also James McKeen Cattell, *Science,* LXIII (1926), 188, for an assertion of U.S. scientific parity with Germany and Great Britain.

Why, then, was this not recognized in 1900? The explanation is that the improvements were not at all obvious at the time. The pay-off on this national investment in research and development did not come immediately or all at once. A time lag was involved. Wasn't it the case, however, that the physical sciences were still conspicuously lagging at this date? A community of physical scientists in 1900, including A. A. Michelson, Simon Newcomb, Henry A. Rowland, J. Willard Gibbs, S. P. Langley, T. W. Richards, G. E. Hale, R. W. Wood, Ira Remsen, and many others of note, was quite respectable, even if not at the top of the heap. Here, too, the full harvest would come in not too many years.

A second proviso relates to technology. The qualitative and quantitative studies previously called for will undoubtedly disclose a technology quite different from the traditional accounts in which inventors loom so large. The Rube Goldberg gadgeteer struggling against the odds in a garret will prove less important than the applier of science and the engineer of systematic production and distribution. What we will need in the future is a study of technologies existing in a domain between the mere empirics of the inventors' world and the aristocratic dispensers of Newton's mechanics.

If the writers on the indifference theme were impelled by beliefs in the overwhelming effects of practical pressures and the pervasiveness of anti-intellectualism—in spite of rather dubious evidence—the views cited here also arise from an intellectual bias. Even in the absence of the requisite evidence, the scientific analogue of the Turner thesis seems dubious. In Turner's view frontier conditions determined a unique American historical development; the indifference theme is an obvious parallel. It is inconceivable that science in America was a unique case because of American conditions. There are too many evidences in print and in manuscript of European-American scientific contacts. If anything, early American science is a splendid example of the opposing historiographical view of the transit of civilization westward—followed in short order by a return flow of research findings.

An early, persistent minor note in the indifference literature and related works on the state of American science is on the paucity of supporting scientific institutions. Complaints of this nature become increasingly rare in this century. The lack of professorial chairs, learned and professional societies, scientific periodicals, university laboratories, and scientific bureaus was once very apparent. America was rich in land, people, and natural resources but poor in institutions. As these were needed to start and to sustain science, the last three-quarters of the nineteenth century in America was a great period of institution-building. Quite often the scientists were consciously trying to create American counterparts of European science —or at least what was believed to be European science. The effort was, by and large, a great success. Not surprisingly, however, this scientific community differed in many respects from the original European model.

Rather than a relatively small group of scientists accorded great esteem, official status, and financial support, the American scientific community started out as an amorphous group struggling to differentiate itself from amateurs. Thanks to the institution-building, amateurs were largely excluded. But American society continued to rate men of practicality as the equals of men of science. As American scientists passionately believed in the importance of their labors, the equation with inventors and the like was a matter of great concern, if not anguish. The anguish possibly arose from unfulfilled aspirations to the supposed social status and power of European scientists in the face of competition with technologists in a presumably anti-intellectual society concerned with getting ahead.

Was science really valued more for itself in Europe? Quite a number of interesting historical studies remain to be done on the American image of scientific Europe and how that image changed as Americans gained more first-hand knowledge. The image was based largely on a few great names—men with seats in ancient universities, men who became noblemen or knights, men who received subventions from the state. The many who could not find positions or even get education at all were forgotten, as was the class structure, which largely determined who entered the circle of science.[29] Gauss

---

29. Cf. Fritz K. Ringer, *Decline of the German Mandarin: The German Academic Community* (Cambridge, Mass., 1969); and *idem,* "Higher Education in Germany in the Nineteenth Century," *Journal of Contemporary History,* II (1967), 123–38; and Cardwell, *Organisation of Science.*

and Faraday were not at all typical. Forgotten also were the abundant signs of European concern with the practical uses of science.[30] Those who praised European support of abstract science often overlooked the degree that support originated in and depended on desires for national prestige and glory for the sovereign. Earning the psychological benefits of prestige and glory are practical goals, just as much as building a better mouse trap or growing two blades of grass where one grew before.

To writers on the indifference theme, it was reprehensible that Americans knew of Thomas Edison but had never heard of his great contemporary, J. Willard Gibbs. Americans *should* know about Gibbs. We could prepare new history textbooks exalting theory over gadgeteering, in which Gibbs looms large and Edison is simply a name in a list of inventors. But why worry about public acclaim if advancement of knowledge is the real concern? Gibbs did much scientific work; loyal to basic research, he was indifferent to mass adulation.[31] There is no point in substituting a pure science myth for an inventor's myth. The man in the street in London is probably not familiar with James Clerk Maxwell's achievements; neither is the French peasantry conscious of Baron Cuvier, nor are there many in the crowds of the Munich Oktoberfest who have ever heard of Karl Friedrich Gauss.

This point illustrates a confusion in the literature about the differences between general or popular knowledge and professional knowledge. In part the confusion stems from the assumption that the two were once identical, i.e., that the knowledge of the public and the knowledge of the working scientist largely coincided. As science became more complex, it is further assumed, the two knowledges diverged and science became the arcane property of an elite. This

---

30. An example of how historiographic commitments affect the secondary literature is Jacob Schmookler, "Catastrophe and Utilitarianism in the Development of Basic Science," *Economics of Research and Development,* ed. R. A. Tybout (Columbus, Ohio, 1962), pp. 19–33. Schmookler has no difficulty, as an economic historian, in finding evidence for applied concerns, even in citing A. R. Hall, a confirmed internalist, to make the point for the period of the scientific revolution of the seventeenth century.

31. See Muriel Rukeyser, *Willard Gibbs* (Garden City, N.Y., 1942); and L. P. Wheeler, *Josiah Willard Gibbs: The History of a Great Mind* (New Haven, Conn., 1951).

thesis is faulty on two grounds: first, in the past most people were not literate and obviously did not partake of whatever science existed; second, even among the educated of the past, it is very doubtful if all could comprehend the full development of theories and facts, even if so inclined. Ptolemy was not exactly easy to comprehend; neither was Copernicus; nor was Newton's mind readily accessible to the public. Obviously, a large portion of the professional scientific knowledge remained professional. What the public should know, and how to reach them, is a major problem, especially in a modern, democratic society.

American scientists looking at Edison in the limelight and Gibbs in solitary splendor at Yale could not overcome their ambivalent feelings about the relations of science to technology. The ties between the two were probably closer than in Europe. The number of scientists was growing, and their institutions were flourishing. Yet the feeling persisted that science was not adequately recognized by society and that the scientists were not honored as such. The scientists did not (and do not) want to be loved for winning the frontier, for advancing technology, for helping to make bombs and missiles, or even for refurbishing ghettos and restoring the purity of the environment, but for advancing knowledge.

The indifference literature is a statement of dismay, maybe despair, at the insensitivity of the American people to the importance of advancing knowledge. The scientists fail to recognize the difference between the American and the European situations. The issue here is not the pressures for practical results or the urge for power and status, although scientists are not immune to these. The main difference is a matter of numbers and the question of how a large, non-elite scientific community can function in a mass culture with an ideology derived from an elitist, European situation.

By 1900, almost all leading American institutions of higher education differed from those in Europe in one basic feature. Instead of one professor for each major field, American institutions had departments with two or more faculty. Instead of a small number of very highly esteemed professors, there were many professors, and therefore more opportunities in science. Only after World War II did some European institutions slowly start departments on the

American model. In America, unlike Europe, science was an activity carried on by a large number of people—not counted in tens or hundreds, but in thousands, tens of thousands, and hundreds of thousands. All were presumably properly educated and formally equal, although in talent and productivity differences, naturally, did exist. All, however, were supposed to do research. Despite recurring complaints about the glut of scientific literature, the number who performed research and the quality of publications are impressive.

The leaders of the scientific community—perhaps the rank and file also—were not always conscious of the change. They were thinking of Europe and looking for a few great men. Two leaders of the American scientific community proposed ways to overcome what they saw as America's scientific deficiencies, especially in the physical sciences. In 1902, R. S. Woodward, who would later become head of the Carnegie Institution of Washington, proposed to its trustees the erection of a laboratory for a few outstanding physicists.[32] George Ellery Hale, in post-World War I planning for the National Academy of Sciences, similarly proposed laboratories for a handful of great investigators.[33] Woodward's proposal was a bit too extravagant for the trustees, who leaned to other fields. Hale's proposal lost to the argument that these few investigators should be at institutions of higher learning. The leaders of the scientific community—at least up to World War II—never seemed to realize the nature of their problem. What was needed was not support for a few great or near-great physicists, zoologists, chemists, physiologists, or mathematicians, but support for the labors of a mass scientific community.

The problem was peculiarly American. English scientists were utterly indifferent to Cockney ignorance of James Clerk Maxwell. After all, only a small percentage of the British population would go to universities. The class structure and the system of interlocking institutions would require that only a small number need know of

---

32. Carnegie Institution of Washington, *Yearbook*, 1902, 13–24.

33. See Ronald C. Tobey, "The New Sciences and Democratic Society: The American Ideology of National Science, 1919–1930" (Ph.D. diss., Cornell University, 1969), pp. 35–94; Daniel Kevles, "George Ellery Hale, the First World War, and the Advancement of Science in America," *Isis*, LIX (1968), 427–37; and Helen Wright, *Explorer of the Universe: A Biography of George Ellery Hale* (New York, 1966).

Maxwell and appreciate science. Nothing like this existed in the America of the last century, already on the way to a mass culture. Edison outpolling Gibbs was bad enough. How could the scientists compete for popular esteem with politicians, show business types, athletes, and evangelists? How could basic research be sold to a society where everyone reads the comics?

To this day, the scientific community is still somewhat ill at ease in a mass culture. It still tries to sell basic research by aligning it with practical goals of popular appeal.[34] More significantly, starting in the last century, the scientific community has fairly consistently attempted—not always successfully—to form privileged sanctuaries for its way of life. Yet the secure position that science now occupies in America is a great credit to the institution-building talents of the scientists. It also indicates that Americans, who may or may not have been indifferent, were not particularly hostile and were remarkably tolerant of this deviant group in their midst.

---

34. Cf. George Daniels, "The Pure-Science Ideal and Democratic Culture," *Science*, CLVI (1967), 1699–1705.

A Statistical Profile
of American Scientists,
1846–1876

In this age of uncertainties and revisionism, even the historians hug the quantifiable. David Donald has quantified abolitionists, George Mowry has quantified Progressives, William Miller has quantified businessmen, and several historians are currently quantifying voters and legislators. American scientists have been analyzed statistically as often as any other class: for example, nineteenth-century scientists by Robert Siegfried, Donald Beaver, and George Daniels, twentieth-century by Robert Knapp and Herbert Goodrich and by Stephen Visher.[1] The group to be considered here consists of 477 American scientists active during any part of the period 1846–76 and distinguished enough to be among the subjects in the *Dictionary of American Biography,* familiarly known as the *DAB.*[2]

The period chosen runs from the outbreak of the Mexican War and consequent sectional hostility through the Civil War and Reconstruction; it might be called the Civil War generation. It also extends from the beginning of one surge in economic growth to the

---

1. Robert Siegfried, "A Study of Chemical Research Publications from the United States before 1880" (Ph.D. diss., University of Wisconsin, 1953); Donald de B. Beaver, "The American Scientific Community, 1800–1860: A Statistical-Historical Study" (Ph.D. diss., Yale University, 1966); George H. Daniels, *American Science in the Age of Jackson* (New York, 1968); Robert H. Knapp and Herbert B. Goodrich, *Origins of American Scientists* (Middletown, Conn., 1952); Stephen S. Visher, *Scientists Starred, 1903–1943, in "American Men of Science"* (Baltimore, Md., 1947).

2. Allen Johnson, Dumas Malone, Harris E. Starr, and Robert Livingston Schuyler, eds., *Dictionary of American Biography,* 22 vols. (New York, 1928–58) (hereafter cited as *DAB*).

end of the next. In American science, the year 1846 was an *annus mirabilis*: Louis Agassiz arrived from Europe, the Smithsonian Institution was founded, the Yale Scientific School was born, that of Harvard twinkled in Abbott Lawrence's eye, and the American Association for the Advancement of Science stirred in the womb. In retrospect American scientists themselves recognized the mid-forties as a watershed. The year 1876 was closely bracketed by the deaths of Agassiz and Joseph Henry, and saw the founding of the American Chemical Society (a sign of deepening specialization) and of the Johns Hopkins University, which, following close upon the innovations of Charles Eliot at Harvard, opened a new age of graduate education in science as in other fields. The period brought the awakening of national scientific ambition and pride, the nationalizing of scientific institutions, the maturing of professionalism, and the advent of the modern American university. It was a period of self-discovery, orientation, and organization—in short, a coming of age.

THE STATISTICAL VALIDITY OF THE DAB

As more than one historian has recognized, the *DAB* is well suited to studies such as this. Financed by the *New York Times,* sponsored by the American Council of Learned Societies, edited by distinguished and dedicated historians, held to high standards of scholarship—its twenty-two volumes include 14,870 articles by 3,052 carefully selected contributors. *DAB* subjects consist of persons who "made some significant contribution to American life in its manifold aspects" and who died before January 1, 1941. The editors' process of selection began with a survey of earlier reference works. Because these sources were found deficient in coverage of scientific and certain other fields, an authority in each of those fields made up a tentative list of subjects, which was circulated for additions and deletions by other specialists in the field. Then necrologies were searched, and the revised and enlarged list was printed and circulated for still more criticism and suggestions. Finally, after the original twenty volumes had been in print for several years, some additional sketches were presented in the first supplementary volume. Some scientists of interest chiefly for their contributions to institutional development, men such as Walter Johnson or Charles Joy, have been left out; but these cases are too few to be statistically significant. *DAB* coverage of scientists is full and fair.[3]

The *DAB*'s index volume, published soon after the original twenty, classifies subjects by occupation, but its analysis for that purpose seldom goes beyond the brief categorization at the beginning of each sketch. A number of subjects with reasonable claims to being scientists were not so listed in that volume. So it was necessary to go through the twenty-two volumes page by page. This resulted in the classification of 477 subjects in from one to three scientific fields each (primary, secondary, and third fields). Each subject was analyzed and classified under twenty-eight additional categories: dates of birth, leaving college, entering primary, secondary, and third fields, period of activity, and death; places of birth, youth (two possible entries), and career (three possible entries); foreign birth of one or both parents; father's occupational status and field; marital status and number of children; religious affiliation; number of years of formal education and (if distinguishable) of formal scientific training; colleges attended; motivations for entering field; age at entry; scientific sources of income and other sources of income (four possible entries for each); length of Civil War service, if any; and impact of the Civil War on career, if determinable.[4] For the sake of consistency, no data were included except those set forth in the *DAB*.

---

3. See *DAB*, I, "Introduction," and XX, "A Brief Account of the Enterprise." As a check, I have reclassified some raw figures of Beaver and Daniels to correspond to my own definition of fields. (Their figures were based on the numbers of published articles and authors in each field.) Among *DAB* scientists active in 1845, my figures show 30.2% of primary fields or 34.6% of combined primary, secondary, and third fields to have been in the earth sciences (see Table 1 for classifications); Daniels shows 28.9% of scientists, 1815–44, and 29.4% of scientists, 1840–45, to have been in that major field; Beaver shows 33.5% of scientists, 1800–1860, and 27.4% of scientists, 1845–49, to have been in it. In the life sciences, my figures for 1845 are 35.1% (primary fields only) and 30.8% (all three fields combined); Daniels' are 32.1% (1815–44) and 32.8% (1840–44); and Beaver's are 29.0% (1800–1860) and 35.8% (1845–49). In astronomy, Daniels' figures are 6–7%, Beaver's 7–8%, and mine 9%. In mathematics, Daniels finds about 3%, I find 8–9%, while Beaver combines mathematics with physics. In chemistry, Beaver's figures range from 15% to 20%, depending on the period, and mine from 15% to 16%, while Daniels combines physics with chemistry. Given the ambiguities of field classifications in that period, the correspondence is generally close.

4. The tabular analysis of biographies was carried out by John B. Cusack.

Two minor distortions were permitted for practical reasons. To minimize subjective judgments, a few subjects were counted as scientists even though they were probably listed in the *DAB* on other grounds. For example, John M. Schofield was counted as a physicist on the strength of his professorship in that field for several years at West Point and at Washington University, his membership in the AAAS, and his efforts to make serious contributions to the field, although his *DAB* sketch classifies him simply as "soldier." Secondly, the number of entrants into science during the seventies may be distorted because a few men who began scientific careers in those years lived past 1940 and so were not included in the *DAB*. But it seems unlikely that either distortion would much affect any other classes of data analyzed here.

The 477 *DAB* subjects analyzed here do not, it is true, include every American scientist or professed scientist active during the period who published a scientific article at any time in his life. Donald Beaver estimates that between 700 and 800 such men were active in 1845, whereas the *DAB* lists only 145 scientists active at that time. But Beaver calculates further that 55% of all published scientific articles by Americans (1800–60) were written by only 17% of the scientists as broadly defined.[5] A reasonable guess, therefore, would be that the 477 *DAB* subjects, as acknowledged leaders, accounted for at least half of American scientific output during the period studied, and the better half at that.

For that reason, it seems immaterial whether or not the statistical profile of this elite group would fit that of the entire body of scientists in this period (though a recent study suggests that, in fact, it would).[6] The results of this analysis are significant in and of themselves—perhaps more significant than would be a profile of the great and little all lumped together. The list of 477 leaders is presented not as a representative sample of some larger group, but as the whole concern of this analysis.

The *DAB* seldom specified religious affiliation. Some other cate-

---

5. Beaver, "Scientific Community," pp. 118, 134.
6. Amos J. Loveday, Jr., "A Statistical Study of the American Scientific Community, 1855–1870" (Master's thesis, Ohio State University, 1970), *passim.*

gories of data, though available, revealed little of significance. But several told much and told it plainly. Tables of these data constitute the core of what follows.[7]

DISTRIBUTION OF SCIENTISTS AMONG SCIENTIFIC FIELDS

Thirty-two per cent of the scientists had a secondary field (one of less importance in the subject's career), and 8% had a third. Among the secondary and third fields combined, physics (12%) and the earth sciences (31%) scored higher than they did among primary fields in Table 1, and the life sciences (23%) registered a lower score; but the other major fields did not vary significantly. Among leading subfields, meteorology, mineralogy, paleontology, and electrical physics scored significantly higher as secondary fields than as primary; and botany, entomology, and natural history scored significantly lower.

Among the scientists with more than one field, the proportion whose secondary and third fields lay outside their primary fields remained constant during the period: 34% of those active in 1846, 35% of those active in 1876, 34% of all counted. This might seem to imply that the degree of specialization did not change. Yet the logic of scientific development and the impressions of contemporaries both argue otherwise. One plausible explanation of the seeming anomaly is the persistent American leaning toward descriptive science, a tendency which continued to encourage unsophisticated descriptive forays into other major fields—by chemists into geology, by mathematicians into geodesy, by astronomers into meteorology, and so on. Another explanation is increased specialization within major fields and subfields, as in the case of the botanist William Sullivant. After 1840, Sullivant began shedding divisions of his subfield—first, grasses and sedges, then lichens and fungi, then all phanerogamous botany, and so on, until by 1860, no longer a rolling stone, he con-

---

7. Most generalizations in my commentary are based on extensive research for a book, tentatively entitled *Science, Union, and Democracy, 1846–1876: The Scientific Enterprise in Civil War America,* to be published by Alfred A. Knopf (probably in 1974). Detailed information would be too voluminous for this essay.

TABLE 1

Primary Major Fields and Leading Subfields, 1846–76

Fields and Subfields	Number of Scientists	% of Total in Major Field	% of All Scientists
Mathematics	41	100.0	8.6
Astronomy	44	100.0	9.2
Earth Sciences [a]	116	100.0	24.4
Geology	60	55.6	
Meteorology	10	9.3	
Geodesy	10	9.3	
Life Sciences [b]	150	100.0	31.5
Botany	51	34.0	
Zoology	26	17.4	
Entomology	25	16.6	
Chemistry [c]	101	100.0	21.1
General	59	58.5	
Industrial	19	18.8	
Agricultural	13	12.9	
Physics [d]	25	100.0	5.2
Mechanical	11	44.0	
Electrical	6	24.0	
Astronomical	4	16.0	

NOTE: Medical sciences were excluded because of the difficulty of drawing lines between theory, technique, and skill in this period. So were 34 *DAB* subjects whose second or third fields were sciences, such as mineralogy, but whose primary fields were technological, such as mining engineering.
[a] Also includes hydrography (8 scientists), oceanography (8), mineralogy (8), geography (8), and topography (4).
[b] Also includes natural history (15), paleontology (12), ornithology (9), conchology (8), ichthyology (2), and anthropology (1).
[c] Also includes analytical chemistry (10).
[d] Also includes spectroscopy (2) and mathematical physics (2).

fined himself to gathering mosses.[8] Such a trend would not show up in this less refined analysis.

If equivalent information existed for scientists of other nations, it would reveal beyond dispute any distinctly American tendency in

---

8. Andrew D. Rodgers III, *"Noble Fellow": William Starling Sullivant* (New York, 1940), p. 244; see Siegfried, "Chemical Research Publications," pp. 105–6, for statistical confirmation of narrowing specialization within chemical subfields (1840–80).

the choice of scientific fields. As it is, however, Table 1 can substantiate the statement of Louis Agassiz in 1847 that "the geologists and mineralogists form the most numerous class among the savants of the country," but not Benjamin Gould's assertion in 1869 that geologists constituted a larger portion of scientists in America than in any other country.[9]

As for the national style of science, contemporaries saw American scientists as tending to aim directly at one small target of opportunity after another, each viewed as an end in itself. This approach was certainly common to Americans generally; it had developed through generations of hastily exploiting the most obvious and accessible natural resources and then evading the law of diminishing returns by moving westward to undepleted territory. In this spirit, American scientists often (as was remarked then and later) looked toward research which promised immediate economic utility. But the contemplated quick return was not necessarily financial; immortality through nomenclature or a journal note was reward enough for some scientists.

Apart from its consonance with American tradition generally, the one-shot, quick-payoff approach was also a natural response to westward expansion, which kept opening up new bonanzas of specimens, formations, and natural phenomena to describe and classify. Doing so might better be called a duty than a fault. American science had its own Manifest Destiny. Still another factor was the circumstance that in America, more than in some other nations, the scientist had to finance his own original research through teaching, routine government work, or sources of income unrelated to science. Such distractions broke up his scientific investigations into spare-time fragments, and so made sustained, elaborately structured researches more difficult.

Thus, during most of this period, American science tended toward accumulating descriptions and measurements rather than building theories, and toward outdoor sciences, like geophysics and natural history, rather than sciences of laboratory and blackboard, such as

---

9. Elizabeth C. Agassiz, ed., *Louis Agassiz: His Life and Correspondence* (Boston, 1890), p. 437; Benjamin A. Gould, Presidential Address, *Proceedings of the American Association for the Advancement of Science*, XVIII (1869), 33.

physics, chemistry, and mathematics. Even among the latter fields, American chemistry seemed more inclined toward description and measurement than theory. Lacking European statistics for comparison, we can only point out that three-fifths of the *DAB* scientists were in the descriptive or outdoor group (including astronomers, whose emphasis in that period was on the measurement and cataloguing of star positions). Even this rough measure must be qualified by the fact that while chemistry, grouped here with theoretical and laboratory sciences, had a strong descriptive flavor at this time, certain earth sciences such as geodesy had a somewhat mathematical cast. Nevertheless, mathematics was used in the earth sciences as a tool of description more than of theory, so the distortion is not gross.

TRENDS IN RECRUITING

Table 2 bears out the contemporary expectation that, with the passing of time and of the frontier, the law of diminishing returns would catch up with descriptive, outdoor researches. Recruitment into the earth sciences fell to a new low percentage in the first half of the seventies, while the combined percentages of chemistry, physics, and mathematics reached a new high. Table 3 shows that the seventies indeed set the pattern of a new era. Though the life sciences, which managed to come indoors to the laboratory, recovered and held their own after 1880, the earth sciences, less adaptable to the new emphasis, continued to decline, while chemistry in the end led all the rest.

In 1968, the National Register of Scientific and Technical Personnel reported statistics on the fields of 242,763 scientists who were registered in 1966. Astronomy was not separately reported. Of the total registered in the five other fields of Table 3, 13.2% were in mathematics, 15.0% in the earth sciences, 17.1% in the life sciences, 38.0% in chemistry, and 16.8% in physics.

The impact of the Civil War is obvious in Table 2. Not only did the war sharply reduce the number of entrants into scientific careers, but, because it postponed the start of college education for younger men, its delayed effect appears in the latter half of the sixties. *DAB* biographies state explicitly that the Civil War directly advanced

TABLE 2

DAB Entrants into Primary Fields by Half-Decades, 1846–75

Period	Total Entrants	% in Mathematics	% in Astronomy	% in Earth Sciences	% in Life Sciences	% in Chemistry	% in Physics
1846–50	47	6.5	10.6	23.5	25.5	30.0	4.3
1851–55	49	8.2	6.1	26.5	32.6	18.4	8.2
1856–60	62	11.4	9.7	19.4	33.9	21.0	4.8
1861–65	41	9.7	9.7	17.0	29.2	29.2	4.9
1866–70	52	3.8	9.6	25.0	32.7	25.0	3.8
1871–75	61	14.7	9.8	16.4	26.1	21.3	11.5
1846–75	312	9.3	9.3	21.1	30.1	23.6	6.4

TABLE 3

Entrants into Primary Fields by Decades, 1881–1940

Period	Total Entrants	% in Mathematics	% in Astronomy	% in Earth Sciences	% in Life Sciences	% in Chemistry	% in Physics
1881–90	508	13.0	3.6	10.5	31.6	26.9	14.0
1891–1900	1,119	16.1	2.9	11.7	37.0	25.2	13.2
1901–10	1,920	9.3	1.9	6.8	36.2	30.8	12.9
1911–20	3,303	5.9	0.7	7.3	39.8	35.2	10.9
1921–30	6,078	5.6	0.7	7.7	36.6	38.2	11.3
1931–40	3,982	5.2	0.8	4.7	32.3	45.6	11.3

SOURCE: Adapted from statistics for mathematics, astronomy, geology, biology, chemistry, and physics in Robert H. Knapp and Herbert B. Goodrich, *Origins of American Scientists* (Middletown, Conn., 1952), p. 324.
NOTE: Totals are of those receiving B.A. degree or equivalent in specified decades and later listed in the third or seventh edition of *American Men of Science* as "starred" (rated highly) or holding Ph.D.'s. The decade 1881–90 is distorted by mortality, since entrants in those years had to live longer into their careers in order to be listed in the editions used, and the decade of 1931–40 is distorted by delay in graduate work because of World War II.

the careers of 10 scientists, in or out of the service, directly injured the careers of 29 (other than by simply interrupting them), and ended the careers of 12. But the *DAB* cannot tell us how many men might have *become DAB* scientists had the war not killed, disabled, or diverted them.

The number of new *DAB* chemists fell only from 13 in the late fifties to 12 in the early sixties, a smaller decline than in any other major field, perhaps because the greater tendency toward European study in chemistry removed more chemistry students from the scene of wartime activities. Measured against numerical strength at the outset, the earth and life scientists exceeded their quotas in Civil War service as reported by the *DAB,* while the other major fields fell short. On the whole, American scientists did their share of service. Only about 240 *DAB* scientists were between fourteen and forty in 1861, yet 231 of them did some military or naval service in the war.

The variations over time in mathematics and (before 1870) in physics, as well as the steadiness of astronomy, involved small numbers of scientists and may have been the result of mere chance. The decline in earth sciences during the late fifties, however, may be explained by a period of hard times in mining as well as by the long-term decline of that industry. The high mark of chemistry in the late forties and its precipitous decline in the early fifties are probably reflections (allowing for an educational lag of three or four years) of the Liebig-induced craze during the early forties for agricultural chemistry as the farmer's cure-all and of the disillusionment that followed in the late forties.

REGIONAL PRODUCTIVITY

From Tables 4 and 5, two facts stand out: the scientific fecundity of New England, especially Massachusetts, and the scientific sterility of the slave states.[10] Table 6 shows that during the period 1846–76

---

10. Cf. Thomas C. Johnson, *Scientific Interests in the Old South* (New York, 1936). Johnson's thesis of Southern scientific parity with other regions in the 1840s and 1850s constitutes a classic example of why quantitative comparisons are desirable.

TABLE 4

DAB Scientists Born Per Million of 1830 Population

Birthplace	Mathematics	Astronomy	Earth Sciences	Life Sciences	Chemistry	Physics	All Fields	N
U.S.	3.0	3.0	7.8	9.9	7.1	1.6	32.3	415
New England [a]	9.2	12.3	18.9	26.6	16.4	3.6	87.0	170
Middle Atlantic [b]	3.4	1.9	10.1	12.1	12.1	2.2	41.8	173
North Central [c]	0.7	2.0	4.8	9.5	2.0	1.4	20.4	30
Border [d]	0.6	1.2	3.0	1.8	0.6	0.0	7.1	12
South Atlantic [e]	1.7	0.3	3.4	1.7	1.4	0.3	9.3	27
Gulf [f]	0.0	0.0	0.0	1.4	1.4	1.4	4.3	3
N	39	38	101	125	91	21	415	

NOTE: 1830 is the census year nearest the median birth date (1831) of the scientists tabulated. No scientists were born in the states omitted below.

[a] New England: Maine, N.H., Vt., Mass., R.I., Conn. [b] Middle Atlantic: N.Y., N.J., Pa., Del., Md., D.C. [c] North Central: Ohio, Ind., Ill., Mich., Wis., Iowa. [d] Border slave states: Ky., Tenn., Mo. [e] South Atlantic: Va., N.C., S.C., Ga., Fla. [f] Gulf: Ala., La. (The 1830 populations of Miss. and Ark. are counted with this region, although no DAB scientists were born in those two states.)

TABLE 5

DAB Scientists Born in Four Leading States (Per Million of 1830 Population)

Birthplace	Mathematics	Astronomy	Earth Sciences	Life Sciences	Chemistry	Physics	All Fields	N
Mass.	14.8	19.7	34.4	52.5	32.8	4.9	159.0	97
Conn.	16.8	16.8	26.8	26.8	13.4	10.1	114.0	33
N.Y.	2.6	2.1	13.6	13.6	11.0	3.1	45.9	88
Pa.	3.7	1.5	10.4	14.1	14.9	2.2	46.7	63

TABLE 6

Birthplaces of Scientists Active in Successive Periods

Birthplace	Active in 1846	Active in 1861	Active in 1876	Entered 1861–76 [a]	All *DAB* Scientists [b]
New England	37.3%	33.8%	33.8%	37.3%	35.6%
Middle Atlantic	34.5	36.8	37.3	36.7	36.3
North Central	0.7	2.2	7.7	13.0	6.3
Border	5.5	2.6	1.9	1.2	2.5
South Atlantic	5.5	8.9	5.6	1.8	5.7
Gulf	0.0	0.4	0.3	0.6	0.4
Europe	16.6	14.1	12.0	8.3	11.8
Other non-U.S.	0.0	1.1	1.3	1.2	1.0
N	145	269	375	169	475

[a] Included to emphasize late trends.
[b] All *DAB* scientists active at any time between 1846 and 1876 whose birthplaces are given in the *DAB*.

New England's phenomenal productivity declined slightly, then rose to an even larger share (given the region's decreased percentage of national population) than at the outset. In contrast, the sharp drop of the South after 1861 (from an already dismal showing) reveals another aspect of the harm done to American science by the Civil War.

The increase in scientists per capita in the North Central states suggests the importance of the duration of settlement; those states increased their share of active *DAB* scientists elevenfold from 1846 to 1876, while from 1820 to 1850 their share of national population increased by only two and a half times. The same factor may help to account for the Middle Atlantic states' maintenance of their proportion of active scientists while their proportion of national population declined from 33% in 1820 to 29% in 1850. Meanwhile the steady decline of Europe's contribution testified to the maturing of American science generally.

New England's share of *DAB* scientists (36%) was consistent with its large share of all *DAB* notables born after 1800. The South, however, got much better marks in all fields combined than in science alone: 20% of the *DAB* notables were born in Maryland, the District of Columbia, Virginia, North Carolina, South Carolina, and Georgia.[11]

A survey of the 250 scientists newly designated as eminent in the 1927 edition of *American Men of Science* showed that New England, on the basis of 1890 population, still outproduced the Middle Atlantic states two to one, the North Central states two to one, the South Atlantic states three to one, and the other Southern states ten to one.[12] For all scientists New England's lead was as great or greater in each case.[13]

A number of factors may be responsible for these geographic disparities. One factor is the association of scientists with cities. Beaver

TABLE 7

Scientists Born in Four Leading States

Birthplace	Active in 1846	Active in 1861	Active in 1876	Entered 1861–76	All *DAB* Scientists
Mass.	20.7%	19.3%	19.2%	21.9%	20.5%
Conn.	9.7	6.7	6.1	5.3	7.1
N.Y.	11.0	18.2	20.8	20.1	18.5
Pa.	17.9	13.4	11.5	11.8	13.2

NOTE: Percentage bases as in Table 6.

finds that 44% of leading American scientists (1800–60) were born in cities, which had only 11% of the nation's 1840 population.[14] For the twentieth century, Visher shows that urban and suburban areas produced a much higher proportion of scientists than the proportion of their population to the national total, and a higher propor-

---

11. Charles O. Paullin, "A Statistical Analysis of the *Dictionary of American Biography*," undated typescript in the *Dictionary of American Biography* MSS., Library of Congress (Washington, D.C.), pp. 13, 20. Paullin's analysis of 13,642 biographies in the first twenty volumes covers nearly every social, economic, geographic, ethnic, and professional aspect of life touched upon by them, and is therefore useful in comparing the statistical profile of leading scientists with that of notable Americans in all pursuits.

12. Stephen S. Visher, *Geography of American Notables* (Bloomington, Ind., 1928), pp. 12–13.

13. Visher, *Scientists Starred*, p. 393.

14. Beaver, "Scientific Community," p. 155.

tion of scientists, moreover, than of notables in all other fields.[15] So, in this study, it is not surprising to find that urbanized Massachusetts yielded 21% of *DAB* scientists as against 12% of *DAB* notables in all fields, New York 19% of scientists as against 16% of all fields, and Pennsylvania 13% as against 9%.[16]

As contemporary scientists often remarked, cities offered better collections of books and specimens, along with the vital stimulus of scientific company. Of 32 American scientific societies in 1846, the New England states had 14, and New York, New Jersey, and Pennsylvania had 11 among them.[17] Most scientists, moreover, were sons of professional men, who were drawn to urban centers. Size alone, however, did not make a city prolific of scientists or even congenial to them. Mid-century scientists complained of loneliness, isolation, and inadequate resources in Cleveland, Cincinnati, San Francisco, and New Orleans. Boston, Philadelphia, and New York, with less than a third of the nation's city-dwellers, produced more than half of its city-born scientific leaders. Even New York's output fell short of its size and activity. About 1845 Boston took the lead from Philadelphia in scientific eminence and held it through the period of this study.[18]

New England's preeminence must have owed something to the fact that the region was an early nucleus of transplanted European culture. Science, being relatively independent of natural resources, tended to be self-recruiting and self-sustaining and thus to endure where first planted firmly. Above all, Yankee forehandedness and strength in public schools, higher education, industrial technology, and learning in general must have figured in the making of Yankee science. Visher and others have further suggested genetic factors, some twists of East Anglian chromosomes, although not claiming to disentangle genetics from cultural inheritance and environment.[19] Even the New England climate may have been physically and intellectually stimulating.

---

15. Visher, *Scientists Starred*, pp. 60–61.
16. Paullin, "Statistical Analysis," pp. 13, 17, 55.
17. Ralph S. Bates, *Scientific Societies in the United States* (New York, 1958), pp. 39, 51, 69.
18. Beaver, "Scientific Community," pp. 157, 359.
19. Visher, *Geography*, p. 137.

A similar multiplicity of factors, in this case adverse ones, may explain the South's backwardness in science. Southern scientists themselves complained bitterly of environmental handicaps: chiggers in the flesh, sweat on the eyepiece, fungus, rot, and insect scavengers among the specimens, and the enervating heat. Several scientists

TABLE 8

Locales of *DAB* Scientists' Activity

	Birthplace [a]	Primary Locale		Secondary Locale	
	Active 1846–76	Active 1846–76[b]	Entered 1861–76[c]	Active 1846–76[d]	Entered 1861–76[e]
New England	41.0%	27.2%	22.9%	20.5%	19.0%
Middle Atlantic	41.6	44.4	47.6	34.5	35.5
North Central	7.2	11.4	15.1	19.5	22.8
Border	2.9	3.3	3.6	6.5	2.5
South Atlantic	6.5	5.4	1.8	8.0	5.1
Gulf	0.7	2.9	2.9	4.0	5.1
Pacific Coast states [f]	0.0	4.5	7.2	3.5	5.1
Rocky Mountain states [g]	0.0	0.9	1.2	3.5	5.1

NOTE: The percentage bases are chosen so as to compare relative ability to produce scientists with relative ability to hold or attract them wherever born.
[a] Birthplaces of 415 native-born *DAB* scientists active 1846–76.
[b] Primary locales of 448 native and foreign-born *DAB* scientists active 1846–76.
[c] Primary locales of 166 *DAB* scientists entering careers 1861–76.
[d] Secondary locales of 200 *DAB* scientists active 1846–76.
[e] Secondary locales of 79 *DAB* scientists entering careers 1861–76.
[f] Pacific Coast states: Calif., Ore., Wash.
[g] Rocky Mountain states: Colo., Wyo.

deputized slaves as collectors in the field—a grotesque and barren forced marriage of oppression to enlightenment. Southern backwardness was somewhat less pronounced in mathematics, in which the physical effort was least. Southerners also kept records of the weather, another activity requiring little exertion. The lack of urbanization (although rural areas elsewhere outdid the whole South in per capita production of scientists), the lack of public schooling, the relative frivolity and unruliness of college students, the low social status of productive physical effort, and the meagerness of libraries and museums all were grave handicaps to Southern science.

Intellectually, as Clement Eaton has suggested, Southern gentlemen lacked the perseverance, concentration, and curiosity necessary for scientific research.[20] They could not stand correction or admit to less than primacy in any circle. A shallow romanticism corrupted their minds. Furthermore, the combination of slaveholding with lip service to both Jefferson and the Bible required a daily exercise in contempt for logic, and the "peculiar institution" aroused hostility between Southerners and intellectual outsiders, including scientists.

Table 8 shows that New England and the South Atlantic states had at least one characteristic in common: an inability to hold or

TABLE 9

Activity in Five Leading Areas

	Birthplace	Primary Locale		Secondary Locale	
	Active 1846–76	Active 1846–76	Entered 1861–76	Active 1846–76	Entered 1861–76
Massachusetts	23.4%	17.6%	15.7%	13.0%	10.1%
Connecticut	8.0	6.5	4.8	3.5	5.1
New York	21.2	14.7	13.3	13.0	13.9
Pennsylvania	15.2	13.4	13.3	4.0	2.5
District of Columbia	0.7	10.3	14.5	13.5	15.2

NOTE: Percentage bases as in Table 8.

attract as many scientists as they produced. For New England, including Massachusetts and Connecticut, the sequence of statistics suggests an acceleration of the "brain drain." This flow was directed largely to the North Central states and the District of Columbia (which, being counted with the Middle Atlantic states, tends to compensate statistically within that region for the losses sustained by

---

20. Clement Eaton, *The Mind of the Old South* (Baton Rouge, La., 1967), pp. 243–44.

New York and Pennsylvania). Before the end of the period, the Far West also began to draw scientists from elsewhere.

SOCIAL AND ECONOMIC ORIGINS

Tables 10 and 11 show the predominance of professional men's sons, both in absolute numbers and relative to their population quota. In 1840 (according to that year's admittedly unreliable census figures), only one in every seventy-five Americans with listed occupations was a professional man; yet three out of five *DAB* scientists

TABLE 10

Status of *DAB* Scientists' Fathers, by Periods of Scientists' Activity

Father's Status	Active 1846	Active 1861	Active 1876	Entered 1861–76	All	N
Professional	58.7%	56.2%	58.9%	64.6%	58.9%	149
Entrepreneur	11.3	18.9	18.3	13.4	17.4	44
Small farmer	16.2	15.0	15.7	15.8	15.4	39
Skilled worker	8.8	5.9	4.5	3.7	5.1	13
Other *	5.0	3.9	2.0	2.4	3.2	8
N	80	153	197	82	253	253

* Five executives, 2 clerks, 1 unskilled laborer.

were sons of professionals. Per capita, the professional group thus produced forty-five times as many scientist sons as did the general population. Indeed, the group's productivity of notable sons was somewhat greater in nineteenth-century science than in all fields and periods combined. Paullin's *DAB* analysis credits them with producing 50% of the total in the latter category.[21] (The figures of Knapp and Goodrich, however, show professionals' sons as constituting only 36% of 583 eminent physicists, chemists, and biologists who graduated from college between the two world wars.)[22]

21. Paullin, "Statistical Analysis," p. 244.
22. Knapp and Goodrich, *Origins*, p. 423.

Explanations for the immense lead of professional men's sons over other classes are obvious. Above all, paternal example and family outlook would predispose a son toward an intellectual, professional career. The thought of such a career might not have even occurred to other young men of that period. Professionals' sons, moreover, were far more likely to attend college and be formally introduced to

TABLE 11

Occupation of *DAB* Scientists' Fathers,
by Periods of Scientists' Activity

Father's Area of Occupation	Active 1846	Active 1861	Active 1876	Entered 1861–76	All	N
Farming	15.8%	18.1%	17.4%	12.3%	17.5%	44
Religion	14.5	11.4	16.9	27.2	16.7	42
Medicine	13.2	10.7	10.3	13.6	12.3	31
Trade and Services	9.2	12.1	12.8	11.1	11.5	29
Education	9.2	12.8	11.3	7.4	9.9	25
Manufacturing	7.9	8.7	8.2	6.2	7.5	19
Law	7.9	6.7	6.7	4.9	6.8	17
Government	10.5	8.7	5.1	1.2	5.6	14
Other [a]	11.8	10.7	11.3	16.1	12.3	31
N	76	149	195	81	252	252

[a] Eight banking and finance, 5 transportation, 5 construction, 4 science, 9 other.

science there. The question of whether or not a college education was "democratic," that is, equally available to all, was much debated in those years. It seems that an ambitious and hardworking young man could usually put himself through college by means of public-school teaching and tuition scholarships, if he did not have a family to support. Even so, this required the sacrifice of immediate returns for a distant—and in that day, not very seductive—goal, a prospect not congenial to most Americans, particularly those low on the income and status scales. Doubtless some movement from nonprofessional classes into science occurred in stages over several generations—for example, from farming to engineering to science. This study does not cover a long enough period to test that hypothesis. But recent studies of such communities as Newburyport, Massachusetts, for the period since the mid-nineteenth century suggest that upward movements in economic and social status were slow and small.

Knapp and Goodrich's figures for a later period suggest the waste of unrealized scientific potential among nonprofessional classes in mid-nineteenth-century America. In contrast to the predominance of professionals' sons, only one of the 253 *DAB* scientists counted was the son of an unskilled laborer. Similarly, Paullin's count of 5,579 *DAB* subjects in all fields lists only 15, or about 1 in 400, as sons of unskilled laborers.[23] Small farmers, who made up 77% of the 1840 population, produced only 15% of the *DAB* scientists (though they fathered 20% of all notables in Paullin's count). Farmers' sons, however, made up 20% of Knapp and Goodrich's 583 twentieth-century scientists, even though farmers had declined to about 17% of the working population by 1910. This increase in spite of diminished numbers may have been the result of farmers' increased prosperity, education, technical expertise, and access to general culture, combined with the development of public colleges and universities and the decline in the number of able young men whom agriculture could absorb.

Throughout the period studied here, as shown in Table 11, the children of educators continued to constitute about a tenth of the *DAB* scientists (as they also did in Knapp and Goodrich's twentieth-century group). The same table shows that the proportion of government officials' and employees' sons declined steadily. The most remarkable change occurred among the sons of clergymen, who after 1861 sharply increased their already impressive proportion to 27% of the new entrants. Perhaps the revelation of Darwinism strengthened an inclination of clergymen's sons to enter a more potent priesthood than the church offered. Of all *DAB* scientists, one in six was a clergyman's son, though clergymen constituted only about one in two hundred of the general working population, a disparity of thirty-three to one. The 1910 census in combination with Knapp

---

23. Paullin, "Statistical Analysis," p. 254.

and Goodrich shows a twenty-seven-to-one disparity still prevailing in the 1920s and the 1930s.

From Tables 12 and 13 it appears that astronomers were more apt to be the sons of small farmers than were other scientists, whereas earth scientists were slightly less likely to be the sons of farmers but significantly more likely to be the sons of clergymen. There is a nice symmetry here, whatever else: sons of farmers looking to heaven, sons of ministers looking to earth. One may easily speculate about that. But why the life sciences and chemistry drew a larger proportion of skilled workers' sons than did other sciences rather baffles speculation, unless they found the craftsmanship of preparing specimens or experiments more congenial. It seems logical, however, that educators' sons should take to the white-collar abstractions of astronomy and mathematics, while physicians' sons leaned toward chemistry. It is also logical that sons of manufacturers should lean toward chemistry and away from the earth sciences. The scarcity of scientists' sons can be explained by the scarcity of scientists in earlier years.

MOTIVATION AND TRAINING

Table 14 shows earth scientists to have been steered toward their fields more than twice as often by jobs as by family influence, whereas no job influence on life scientists was found, but the family influence was important. An obvious explanation is the existence at that time of nonscientific jobs in enterprises related to the earth sciences, especially mining, whereas no such jobs had yet grown from the life sciences. A century later, formal education had superseded on-the-job training, and so by then the job influence had become negligible in recruiting to science, except in chemistry, where it was only about 4%. At the beginning of the 1920s, the influence of school or college still accounted for about half the scientists, but during the next two decades the figure slipped to about 40%. Meanwhile, after declining to about 15% for the second quarter of the twentieth century, family influence began rising again, along with the influence of independent reading. Presumably the growth of library facilities

TABLE 12

Status of *DAB* Scientists' Fathers, by Major Fields of Science

Father's Status	Mathematics	Astronomy	Earth Sciences	Life Sciences	Chemistry	Physics	All	N
Professional	59.0%	50.0%	70.1%	54.5%	57.4%	63.9%	58.9%	149
Entrepreneur	13.6	16.7	16.7	19.3	18.5	9.1	17.4	44
Small farmer	18.2	25.0	11.1	14.8	16.7	9.1	15.4	39
Skilled worker	4.5	4.2	1.9	6.8	7.4	0.0	5.1	13
Other [a]	4.5	4.2	0.0	4.6	0.0	18.2	3.2	8
N	22	24	54	88	54	11	253	253

[a] Five executives, 2 clerks, 1 unskilled laborer.

TABLE 13

Occupation of *DAB* Scientists' Fathers, by Major Fields of Science

Father's Area of Occupation	Mathematics	Astronomy	Earth Sciences	Life Sciences	Chemistry	Physics	All	N
Farming	18.2%	25.0%	13.0%	19.3%	17.0%	20.0%	17.5%	44
Religion	13.6	12.5	22.2	14.8	15.1	30.0	16.7	42
Medicine	4.5	4.2	14.8	13.6	15.1	10.0	12.3	31
Trade and Services	9.1	8.3	11.1	12.5	13.2	10.0	11.5	29
Education	22.7	25.0	9.3	4.5	5.7	20.0	9.9	25
Manufacturing	4.5	4.2	1.9	10.2	13.2	0.0	7.5	19
Law	9.1	4.2	9.3	4.5	9.4	0.0	6.8	17
Government	4.5	0.0	7.4	9.1	1.9	0.0	5.6	14
Other [a]	13.6	16.7	11.1	11.4	9.4	10.0	12.3	31
Total Number	22	24	54	88	53	10	252	252

[a] Eight banking and finance, 5 transportation, 5 construction, 4 science, 9 other.

Table 14

Influences on Choices of Career by *DAB* Scientists Active 1846–76

	Mathematics	Astronomy	Earth Sciences	Life Sciences	Chemistry	Physics	All Fields
School	75.0%	36.4%	51.6%	51.1%	54.8%	40.0%	52.0%
Family	25.0	27.3	9.7	35.6	19.4	40.0	22.9
Job	0.0	18.2	22.6	0.0	6.5	0.0	9.9
Other	0.0	18.2	16.1	13.3	19.4	20.0	15.3
N	8	11	31	45	31	5	131

and mass communication was taking over some of the work that schools and jobs had once done in stirring young minds to science.[24]

Of the 145 *DAB* scientists active in 1846, only one in five is described as having gone through a formal course of scientific or technological study (as distinct from a regular college curriculum). Yet science was a learned profession. So even among scientists born in the 1790s, four out of five in Beaver's tally had some college education, and the proportion rose steadily to 96% of those born in the 1820s.[25] Of the 117 active scientists in 1846 whose education is specified by the *DAB* (most born earlier than the 1820s), 91% attended college and 79% graduated. Among the 413 counted for the whole period, the number of college graduates was 83%.

This does not mean that every graduate who became a scientist had gone to college for that purpose. Some may have done so in order to prepare for general college teaching and, like the elder Benjamin Silliman, had happened to end up teaching science. Others may have been attracted to science after encountering it in college. More fundamentally, perhaps, the sort of young man who, from temperament or background, was likely to enter science was also the sort likely to go to college. Cause, result, or cognate, the association of scientific careers with college degrees is clear.

The most conspicuous change during the period came in graduate study. The proportion who earned degrees beyond the bachelor's level increased substantially, but even more marked was the shift from the M.D. to the Ph.D. A third of all Beaver's scientists and fully three-fifths of his life scientists had M.D. degrees.[26] In the 1846 group of *DAB* scientists, the M.D. was still held by one in four. But over the next thirty years, as specialization divided physicians more and more sharply from scientists, the latter, especially the chemists and physicists, turned instead to the Ph.D. (earned mostly in Europe during this period). In those years more Ph.D.'s than M.D.'s joined the ranks of *DAB* scientists (44 as against 32). A study of 85

---

24. Knapp and Goodrich, *Origins,* p. 423.
25. Beaver, "Scientific Community," p. 358.
26. *Ibid.,* pp. 162, 357.

TABLE 15

Extent of Formal Education Among *DAB* Scientists Active 1846–76

Level of education	Mathematics	Astronomy	Earth Sciences	Life Sciences [a]	Chemistry	Physics	All Fields	N
No college	2.5%	7.5%	7.5%	9.6%	3.2%	9.5%	6.8%	28
College (no degree)	0.0	7.5	11.8	12.0	11.7	9.5	10.1	42
Bachelor's degree	75.0	65.0	58.0	43.3	41.5	47.6	51.5	213
Master's degree	12.5	12.5	4.3	4.0	3.2	4.8	5.6	23
Ph.D.	10.0	2.5	8.6	6.4	23.4	19.1	11.4	47
M.D.	0.0	5.0	9.7	24.8	17.0	9.5	14.5	60
N	40	40	93	125	94	21	413	413

NOTE: Each scientist is counted in only one educational category, the highest he achieved.
[a] Excludes one scientist reported to have had no formal education.

geologists with the M.D. degree shows that more of them received that degree in the 1830s than in any other decade and also shows a sharp drop in recipients after about 1850.[27]

The results tabulated in Table 16 are not surprising. As might be expected, New England colleges led those of other regions by wide margins in the production of scientists, both overall and in each major field except physics.

The proportion of mathematicians among scientific alumni was higher at West Point (29%) than at any other college except Columbia. The French influence at West Point, and perhaps an intellectual kinship between mathematics and military strategy, make this understandable. With Benjamin Peirce on its faculty, Harvard graduated a higher proportion of notable mathematicians than most other schools and led in absolute numbers. Harvard, Michigan, and Dartmouth, all of which had good observatories with able directors, were strong in astronomy. Yale's early lead in astronomy had been lost after the 1830s for lack of facilities and administration support. But Yale dominated the earth sciences, proportionally and absolutely, with the two Benjamin Sillimans, Denison Olmsted, Elias Loomis, George Brush, and James Dana.

In the life sciences, rather incongruously, West Point had the highest proportion of notables among its alumni—perhaps through the influence of Jacob Bailey, the microscopist. West Point was followed in this respect by the University of Pennsylvania with its strong medical tradition and its access to major natural history collections in the Philadelphia societies. In absolute numbers, however, Harvard led in the life sciences with the attraction of Agassiz and his museum; next, in absolute terms, was Yale with its strength in paleontology.

In chemistry, the University of Pennsylvania profited from a strong early start with Robert Hare and from Philadelphia's preeminence in pharmaceutical and industrial chemistry. Yale, with the Sillimans, John Norton, John Porter, and Samuel Johnson, was second to Pennsylvania in the proportion of chemists among its *DAB*

---

27. William Browning, "The Relation of Physicians to Early American Geology," *Annals of Medical History*, III (1931), 547–60, 565.

alumni, and in absolute terms tied Harvard College for first place. In physics, no clear pattern had emerged by 1876.

THE ECONOMIC FRAMEWORK

In Table 17 we see the passing of the amateur element in American science, as revealed by the near disappearance of scientists with no scientific income at all and the decline in the number who supplemented their income through nonscientific work. The steadiness of the proportion of scientists in federal employ (Table 17) and the shift of scientists to Washington (Table 9) indicate a tendency to centralize the federal scientific establishment rather than to enlarge its relative weight in the scientific community.

It may be of interest to note that in 1966, among all 242,000 American professional scientists, 36% worked for educational institutions, 34% for industry and business, 10% for the federal government, 3% for state and local governments, and the balance in miscellaneous situations.[28] This suggests that industry has tended to displace government as an employer of scientists. But these figures apply to all working scientists, not an elite; and, more important, they take no account of federal influence through contracts with both universities and industry.

These tables could be used to reconstruct the typical mid-nineteenth-century *DAB* scientist, defined as one conforming to the most numerous element in each statistical category. He would be the son of a professional man, specifically a clergyman. He would have been born in Massachusetts during the 1830s, would have received a bachelor's degree from Harvard, would have served in the Civil War, and, finally, having been led into geology by school influences, would have taught that subject in a Massachusetts college.

If Charles H. Hitchcock (1836–1919) had only taken his cue from school instead of his father, had gone to Harvard instead of Amherst, and had done the patriotic thing in 1861, he would have been the complete "type specimen." But Hitchcock chose to do none

---

28. National Science Foundation, National Register of Scientific and Technical Personnel, *Summary of American Science Manpower, 1966* (Washington, D.C., 1968), p. 1.

Table 16

American Colleges Attended by DAB Scientists

	Mathematics	Astronomy	Earth Sciences	Life Sciences	Chemistry	Physics	Total
New England:							
Harvard	8	8	7	19	13	3	58
Lawrence Scientific School[a]	2	0	2	2	2	0	8
Yale	7	5	14	13	13	2	54
Amherst	0	1	3	2	0	0	6
Dartmouth	1	4	1	0	0	0	6
Other	2	1	5	15	8	2	33
Total	20	19	32	51	36	7	165
Middle Atlantic:							
University of Pennsylvania	0	0	4	12	12	0	28
West Point	6	0	2	10	1	2	21
Rensselaer Polytechnic Institute	1	0	2	0	3	2	8
Columbia	3	0	1	1	1	0	6
Union College	0	2	1	2	1	0	6
Other	3	4	15	20	8	5	55
Total	13	6	25	45	26	9	124

	Mathematics	Astronomy	Earth Sciences	Life Sciences	Chemistry	Physics	Total
North Central: University of Michigan	0	5	2	1	3	0	11
Other	0	2	6	9	2	1	20
Total	0	7	8	10	5	1	31
South	3	1	4	2	5	1	16
U.S. Total	36	33	69	108	72	18	336

NOTE: Only one college, the one he attended longest or from which he received his first degree, was recorded for each scientist.
* Affiliated with Harvard.

of those things, and no one else came even as close as he did to the statistical ideal. The untidy-minded who like to see statistics confounded, a class which includes or at least ought to include most

TABLE 17

Scientific Sources of *DAB* Scientists' Income

	Active 1846	Active 1861	Active 1876	Entered 1861–76	All [a]
Education	39.9%	43.1%	45.4%	48.6%	44.5%
U.S. government	22.8	22.3	22.0	21.1	21.7
State government	9.9	9.7	11.1	11.5	10.5
Wages and salary	5.7	5.4	5.6	7.0	6.3
Entrepreneurship [b]	3.6	5.1	4.9	3.3	4.4
Fees [c]	3.6	3.8	4.5	4.8	4.2
Royalties	4.1	4.0	3.8	3.0	3.8
Other	0.0	0.3	0.4	0.4	0.3
None [d]	10.4	6.4	2.4	0.4	4.2
Number of sources [e]	193	373	551	270	683
Number of scientists	137	252	359	175	452
% *DAB* scientists with nonscientific income [f]	41.4%	40.1%	29.9%	20.7%	32.8%

[a] 452 scientists active at any time between 1846 and 1876, and having sources of income specified in *DAB*.
[b] E.g., chemical, optical, electrical enterprises.
[c] E.g., medical fees, consulting fees.
[d] I.e., amateur scientists.
[e] As many as four sources could be listed for each scientist.
[f] Percentage of all *DAB* scientists active in the given period with or without specified sources of income.

historians, owe the late Professor Hitchcock a vote of thanks for thus having dodged Harvard, the draft, and the passionless punch-card. He and his still more aberrant brethren remind us that a statistical average, however suggestive of historical reality, is not necessarily identical with it.

The Political Economy
of Science

On May 23, 1842, a large group of interested citizens assembled in Cincinnati, Ohio, to organize an astronomical society. The prime mover in the enterprise was Ormsby MacKnight Mitchel, a local savant whose lantern-slide lectures on popular astronomy had captured the community's imagination. Fired by his enthusiasm and determined to establish, on the basis of public subscription, the finest observatory in the United States, the group met to draft a constitution and by-laws. The document was of more than passing interest. Hidden away in the grandiloquent phrasing of its preamble was a pattern of attitudes and a set of principles which ranged far beyond the local concerns of a culture-conscious river town in the 1840s.

Believing it to be the duty of every people, to foster science, . . . recognizing to the fullest extent, the claims which the world has had on the several republics, composing the United States, to contribute to the promotion of science . . . ; realizing the truth, that in our own country, and under a republican form of government, the people must hold, with respect to all great scientific enterprises, the position of patrons, which in monarchical governments is held by Kings and Emperors; and knowing that our country is comparatively deficient in means and instruments, to accomplish original observations in astronomy; Therefore, for the purpose of furnishing our city with an Observatory and Astronomical instruments, in all respects adequate to the wants of science, we, the undersigned, have united, and agree to contribute the sums already subscribed.[1]

---

1. John Quincy Adams, *An Oration Delivered before the Cincinnati Astronomical Society, on the Occasion of Laying the Corner Stone of an Astronomical Observatory, on the 10th of November, 1843* (Cincinnati, Ohio, 1843), Appendix; *Siderial Messenger,* I (July, 1846), 1–2, and (August, 1846), 9–11.

The objectives of the Cincinnati Astronomical Society were limited, immediate, and practical. But in setting them down the members unconsciously had certified a fact, revealed an attitude, and hinted at a problem, all of which explained a great deal about science in nineteenth-century America. The fact was that ordinary Americans were not necessarily indifferent to pure science. The attitude was that private initiative and voluntary association were particularly appropriate methods of funding science in a democratic republic. The problem was the hiatus between the private and public sectors in this regard. These principles and attitudes, defined and expressed primarily through action rather than through self-conscious theorizing, governed the political economy of science in nineteenth-century America.

THE ILLUSION OF LAISSEZ FAIRE

In some circles it was, and is, fashionable to regard the nineteenth century as the Golden Age of laissez faire and sturdy individualism. According to whiggish moral philosophers, Gilded Age jurists, and self-effacing millionaires, during those years Adam Smith's "obvious and simple system of natural liberty" held sway. Horatio Alger stood as the living embodiment of civic virtue. Late in the century, however, government began to encroach on what had been exclusive spheres of private action. This shift in public policy generated a serious ideological conflict between champions of laissez faire and promoters of the general welfare state.[2]

Such a formulation is too tidy and too abstract. It may explain something about the dialectics of political philosophy, but it explains very little about nineteenth-century American life. Blessed by seemingly unlimited natural resources, and by widespread individual

---

2. Joseph Dorfman, *The Economic Mind in American Civilization,* 5 vols. (New York, 1946–59), Vols. II, III; Sidney Fine, *Laissez-Faire and the General Welfare State* (Ann Arbor, Mich., 1956), *passim;* James Willard Hurst, *Law and the Conditions of Freedom in the Nineteenth Century United States* (Madison, Wis., 1956), *passim;* Yehoshua Arieli, *Individualism and Nationalism in American Ideology* (Cambridge, Mass., 1964), *passim;* Max Lerner, "The Triumph of Laissez-Faire," in *Paths of American Thought,* ed. Arthur Schlesinger, Jr., and Morton White (Boston, 1963), pp. 147–66.

opportunity to exploit them, Americans were often in too much of a hurry to maintain nice distinctions between public and private enterprise. "Everyone is in motion, some in quest of power, others of gain," wrote Alexis de Tocqueville, speaking for a host of other foreign observers. "Everything whirls around, . . . and man himself is swept and beaten onwards by the heady current."[3] In such a setting, the public-private dualism had little real vitality apart from its relation to substantive issues rooted in time and place.

Moreover, the record of what Americans did and said makes it clear that only a few purists believed in the self-denying maxim that that government is best which governs least. Their conception of laissez faire was positive, not negative; the proper role of government was not to sit idly by, but rather to act affirmatively to enlarge and expedite the applications of individual and social energy. As Simon Newcomb explained it, there was a difference between the let-alone principle, which he approved, and the keep-out principle, which he did not. "The one claims that the government should not stop the citizen from acting; the other that it should not act itself."[4]

Even the classical economists had allowed considerable public initiative, advancing laissez faire more as a prudent policy than as an iron law. Adam Smith sanctioned state aid for essential social services, and for other institutions which were not sufficiently profitable to attract private enterprise. Jean-Baptiste Say, whose *Treatise on Political Economy* was the most widely used economics text in ante-bellum colleges, specifically recommended subsidies for science, especially for those investigations "of no immediate or apparent utility." John Stuart Mill, whose political economy also enjoyed considerable vogue, agreed that science was too vital to be left to the uncertain stimulation of supply and demand. Indeed, when it came to the subvention of scientific research, the classical economists were surprisingly close to the mixed mercantilism of Alexander Hamilton. "In countries where there is great private wealth," he had written, "much may be effected by the voluntary contributions of patriotic

---

3. Alexis de Tocqueville, *Democracy in America,* 2 vols. (New York, 1945), II, 43.
4. Simon Newcomb, *Principles of Political Economy* (New York, 1885), pp. 443–44, 452–56.

individuals; but in a community situated like that of the United States, the public purse must supply the deficiency of private resources." [5]

Obviously, not all Americans shared Hamilton's high federalism, nor could they live with his easy reconciliation of public and private enterprise. During the early years of the Republic, science policy— like public policy in general—groped for constitutional foundations. A significant number of the Founding Fathers were men of science, if only by avocation, and strict constructionists, if only by profession. No one knew exactly what the commerce and general welfare clauses might promise for federal science, nor were there clear guidelines indicating what the government might, or ought to, do in the interest of internal improvements or standardized weights and measures. Several decades of intermittent debate over the propriety of national universities, academies, and observatories produced neither a clear body of doctrine nor a centralized scientific agency. Those years did, however, produce a policy of sorts: an ad hoc American system of scientific enterprise which responded pragmatically to intellectual demand and political pressure, and which respected, sidestepped, or quietly ignored constitutional scruples as the occasion required.[6]

Doctrinaire strict constructionism was not, as a result, a serious deterrent to government science in nineteenth-century America. Objections based on constitutional principle were usually halfhearted, or derived their strength from association with more potent arguments. Opponents of the Wilkes Expedition, for example, employed ridicule more often than constitutional law. The enterprise was just another "chimeral and harebrained" idea, they charged, as nonsensical as John Quincy Adams' observatory or John Symmes's idea of a hollow earth. Next Congress would be asked to send an expedition

---

5. Adam Smith, *An Inquiry into the Nature and Causes of the Wealth of Nations,* 6th ed., 2 vols. (London, 1899), II, 241, 341; Jean-Baptiste Say, *A Treatise on Political Economy,* 4th ed. (Philadelphia, Pa., 1844), pp. 201–2; John Stuart Mill, *Principles of Political Economy,* 5th ed., 2 vols. (New York, 1920), II, 569–72; William Graham Sumner, "Laissez-Faire," in *Essays of William Graham Sumner,* ed. Albert Keller and Maurice Davie, 2 vols. (New Haven, Conn., 1934), II, 472; Jacob E. Cooke, ed., *The Reports of Alexander Hamilton* (New York, 1964), p. 204.

6. A. Hunter Dupree, *Science in the Federal Government* (Cambridge, Mass., 1957), pp. 1–65.

to the moon, or to underwrite "the plan of the gentleman from Massachusetts, for erecting light-houses in the skies."[7]

Similarly, constitutional reservations about accepting James Smithson's bequest for a scientific establishment in the nation's capital were not directed against the impropriety of government science as such. John C. Calhoun and other Southerners convinced themselves that a new institution in Washington would pose an added threat to state sovereignty.[8] Thomas Hart Benton and John Chipman feared it would grow into another corporate monster like Biddle's bank. What distinction was there, Chipman demanded, "between a corporation in the form of a United States Bank and a corporation intended to elevate humanity"? Indeed, "a corporation by any other name should be as offensive to the democracy."[9] These were fighting words in the Age of Jackson, and it is significant that they carried little if any weight. By the time Joseph Henry took charge of the Smithsonian Institution, the federal government was already deeply committed to a variety of exploring expeditions on land and sea, to the Coast Survey, the Patent Office, the Topographical Engineers, and to a Naval Observatory thinly disguised as a Depot of Charts and Instruments. And science would prove hard to get out once the government had let it get in.[10]

On the state level the call for public science sounded along with the clamor for canals, railroads, and other internal improvements. The decidedly mercantilist tone of early-nineteenth-century state governments silenced whatever laissez faire objections there might otherwise have been to the funding of research from that quarter.[11]

---

7. David P. Tyler, *The Wilkes Expedition* (Philadelphia, 1968), pp. 6, 23; U.S. Congress, House, *Congressional Register,* 24th Cong., 1st sess., May 5, May 9, 1836, pp. 3472, 3554, 3565, 3578; House, *Congressional Globe,* 25th Cong., 2d sess., April 9, 1838, p. 273.

8. William H. Rhees, ed., *The Smithsonian Institution: Documents Relative to Its Origin and History, 1835–1899,* 2 vols. (Washington, D.C., 1901), I, 173–76, 350–51.

9. *Ibid.,* I, 179, 313, 391.

10. Dupree, *Science in the Federal Government,* pp. 44–65.

11. Robert A. Lively, "The American System: A Review Article," *Business History Review,* XXIX (March, 1955), 81–96; Joseph Spengler, "Laissez-Faire and Intervention: A Potential Source of Historical Error," *Journal of Political Economy,* LVII (October, 1949), 438–41; Henry A. James, "Private Corporations and the State," *Journal of Social Science,* XXIII (November, 1887), 145–66; Hurst, *Law and the Conditions of Freedom,* pp. 1–70.

The state geological surveys began with Amos Eaton's efforts in New York in 1821–22, which were privately financed by his patron, Stephen Van Renssalaer. The following year Denison Olmsted, a Silliman student from Yale, convinced the North Carolina legislature that a geological survey would simultaneously advance the economic interest of the state, the science of geology, and his own professional career. From New Haven, Benjamin Silliman offered the imprimatur of the *American Journal of Science*. "We cannot doubt, that (*if adequately encouraged by the local government, or by patriotic individuals,*) the enterprise will produce important advantages to science, agriculture, and other useful arts."[12] Following North Carolina's lead, in the 1830s and 1840s twenty other states established surveys. By 1900, a total of thirty-three states had invested nearly $6 million in geological and geographical reconnaissance.[13]

Naturally, many of the state legislatures intended the surveys to be little more than prospecting expeditions, undertaken at public expense for the benefit of private capitalism. But all the surveys required the services of competent specialists, and, like the Wilkes Expedition, the state surveys provided an entire generation of scientists with employment and experience in a number of critical fields. Some, such as Edward Hitchcock in Massachusetts and David Dale Owen in Indiana, conscientiously tried to give the practical-minded taxpayer his money's worth. Others, notably James Hall in New York and the Rogers brothers in Pennsylvania, scarcely attempted to mask their abstruse researches under utilitarian guise. When the surveys came under attack, as they frequently did, it was because of hostility to pure science, not hostility to state science.[14]

Thus Benjamin Silliman's parenthetical remark about the North

---

12. *American Journal of Science*, V (1882), 202; Walter B. Hendrickson, "Nineteenth Century State Geological Surveys: Early Government Support of Science," *Isis*, LII (September, 1961), 358–59.

13. George P. Merrill, "Contributions to a History of American State Geological and Natural History Surveys," United States National Museum, Bulletin 109 (Washington, D.C., 1920), pp. 537–38.

14. Hendrickson, "Nineteenth Century State Geological Surveys," pp. 361–71; Gerald D. Nash, "The Conflict between Pure and Applied Science in Nineteenth Century Public Policy: The California State Geological Survey, 1860–1874," *Isis*, LIV (June, 1963), 217–28.

Carolina survey, that it would succeed *"if adequately encouraged by the local government, or by patriotic individuals,"* had broader implications. Like most scientists, Silliman cared little where the money came from, so long as it came. Nineteenth-century American scientists were primarily concerned with two goals which were to them inseparable: the advancement of pure science and the practical advancement of their own professional careers. Fresh from an inspirational tour of Europe, Joseph Henry issued the classic call to action in 1838: "The real working men in the way of science in this country should make common cause and endeavour by every proper means unitedly to raise their own scientific character. To make science more respected at home to increase the facilities of scientific investigations and the inducements to scientific labors."[15] Henry's ambitious program left little time to ponder abstract problems of political economy.

SUBSIDIZED PROFESSIONALIZATION

What *did* concern Henry's generation, and later generations, were logistical problems stemming in part from the very advancement of science itself. From the Baconian era until Henry's time, original research could be a part of almost any educated person's experience. The learned man was likely to be also an active physician, merchant, politician, or minister. But as such men busily added to the stock of scientific knowledge, they made it increasingly difficult and finally impossible for subsequent generations to maintain that style of scientific enterprise. Thomas Jefferson succeeded; John Quincy Adams tried and failed; Old Hickory didn't try. University training replaced self-improvement; the technical journal supplanted the gentleman's magazine; science left the parlor cabinet and moved into the lab. In effect, as science became more complex and expensive, it also became increasingly remote from the experience of those laymen who were asked to finance it. Scientists found them-

---

15. Henry to Alexander Dallas Bache, August 9, 1838, Joseph Henry Papers, Smithsonian Institution, Washington, D.C.; George H. Daniels, "The Process of Professionalization in American Science: The Emergent Period, 1820–1860," *Isis,* LVII (Summer, 1967), 151–66.

selves caught between their allegiance to the pure science ideal and their practical realization that to do research at all they might have to court Philistine politicians and millionaires. A Louis Agassiz or a George Ellery Hale could live in both worlds with little apparent discomfort. Others, like Agassiz's son Alexander, or the botanist George Engelmann, never reconciled themselves to a beggar's role. "I do not understand the soft soaping as writers phrase it," Engelmann complained in 1860. "A man who has no real scientific zeal nor knowledge who must be got to do things by diplomacy, I can not do much with." [16]

In the main, however, the scientific community was not selective about its patronage. It sought and accepted funding from any quarter with few questions asked. Joseph Henry treated the peculiar public/ private character of the Smithsonian as a circumstance to be exploited rather than as a problem to be solved. Silliman's son Benjamin, John P. Norton, and John A. Porter parlayed a local interest in agricultural chemistry into the Sheffield Scientific School —with a little help from Porter's wealthy father-in-law. Louis Agassiz, easily the champion grantsman of nineteenth-century American science, impartially welcomed private contributions, appropriations from the Commonwealth of Massachusetts, and preferential treatment from the United States Coast Survey. A. A. Michelson worked for the government's Naval Observatory and for John D. Rockefeller's Standard Oil University. A New York botanist summed up the situation concisely at a fund-raising banquet in 1892: "It matters little whether the support of the university or of special institutions for research comes from Government or from private endowment, provided the provision is adequate and constant." [17]

The funding of nineteenth-century American science was never constant, nor from the scientists' point of view was it ever adequate. They frequently bemoaned America's inferior position in the sci-

---

16. Engelmann to Asa Gray, April 10, June 12, November 1, 1860, Historical Letter File, Gray Herbarium, Harvard University, Cambridge, Mass.

17. Addison Brown, "Endowment for Scientific Research and Publication," *Smithsonian Annual Report, 1892* (Washington, D.C., 1893), p. 627; Howard S. Miller, *Dollars for Research: Science and Its Patrons in Nineteenth Century America* (Seattle, Wash., 1970), *passim.*

entific world, hoping to convince potential patrons that the cause of science was the cause of social progress and national prestige. The general public did still tend to confuse teaching with research, and applied science with pure science. Most scientists, as a consequence, most readily found employment either in the universities or in government bureaus. By the end of the century, the total private endowment specifically earmarked for research was less than $3 million.[18]

Nineteenth-century scientists and some twentieth-century historians focused on the negative picture, and soothed their frustrations by denouncing American indifference to basic science. Such a view overlooked the substantial number of universities, laboratories, scientific schools, and research agencies which had been founded since the 1830s. It also obscured the fact that, in an era before modern research and development had made science too obvious a feature of daily life to ignore, it was asking a great deal to expect a layman to share the values of a professional research scientist. The remarkable thing was not that scientists lacked support, but that they received the support they did.

SCIENCE POLICY IN A CORPORATE AGE

Patterns of support tended to follow general trends in social and economic development. During the second half of the century, as corporate capitalism rationalized business enterprise and finance, the funding of scientific research slowly evolved from occasional, piecemeal patronage to million-dollar foundations and broad-ranging, multifaceted scientific bureaus. Gilded Age folklore notwithstand-

---

18. George Featherstonhaugh, "On the Comparative Encouragement Given to the Study of Natural History in Europe and North America," *The Monthly American Journal of Geology and Natural Science*, I (March, 1832), 391–407; Henry Rowland, "A Plea for Pure Science," *Proceedings of the American Association for the Advancement of Science*, 1873 (Salem, Mass., 1884), 104–22; Carl Snyder, "America's Inferior Position in the Scientific World," *North American Review*, CLXXIV (January, 1902), 59–72; Simon Newcomb, "Conditions Which Discourage Scientific Work in America," *ibid.* (February, 1902), pp. 145–58. Cf. Richard Shryock, "American Indifference to Basic Research during the Nineteenth Century," *Archives internationales d'histoire des sciences*, XXVIII (1948), 50–65; I. Bernard Cohen, "Science in America: The Nineteenth Century," in Schlesinger and White, *Paths of American Thought*, pp. 167–89.

ing, these transitions were not seriously hampered by the ideological counterclaims of public versus private enterprise. It was true, as James Bryce noted in the 1880s, that Americans paid lip service to laissez faire. But their everyday behavior showed little more regard for economic orthodoxy than it had decades earlier. "Economic theory did not stop them," Bryce wrote, "for practical men are proud of getting on without theory. The sentiment of individualism did not stop them, because state intervention has usually taken the form of helping the greater number, while restraining the few." Bryce believed that the social effects of science and technology had shown Americans "how many things may be accomplished by the application of collective skill and large funds which are beyond the reach of individual effort." [19]

The Gilded Age surveys of the western territories were a prime example of what collective skills and large funds could do. The scientific results were impressive, particularly in vertebrate paleontology. But the surveys had evolved competitively during a period of salutary congressional neglect. They had done their work at the cost of administrative chaos in Washington and massive duplications of effort in the field. The reckoning came in 1878, when Congress asked the National Academy of Sciences to recommend a scheme for consolidating the six existing agencies into a single, more efficient, and less expensive geological survey.

The NAS report and the subsequent legislation touched off a controversy over science in the federal government that sputtered on for nearly a decade. Many of the scientists involved, and some historians since, mistook the imbroglio for a turning point in the history of American science. It was not. In the end, the Allison Commission merely ratified, formally, the status quo. The war of words in the 1880s was significant not because it introduced new policies, but because it legitimated old practices. Most of the rhetoric

---

19. James Bryce, *The American Commonwealth*, rev. ed., 2 vols. (New York, 1924), II, 407, 593–94. See also Albert Shaw, "The American State and the American Man," *Contemporary Review*, LI (May, 1887), 698; Fine, *Laissez-Faire and the General Welfare State*, pp. 30–164; Hurst, *Law and the Conditions of Freedom*, pp. 71–108.

only thinly disguised partisan political opportunism, while those few individuals who did sincerely stand on principle were powerless to influence the events that swirled around them.[20]

Few members of Congress realized that the 1878 call for a consolidated survey was more than a straightforward economy move. It was also the first phase of Clarence King's drive to establish a centralized, even more ambitious, federal survey under his own direction. King's strategists purposely kept the issues clouded by advertising retrenchment as a vindication of laissez faire. Survey reform, said John Wesley Powell, would confirm the principle that government science should be "very limited and scrupulously defined."[21] James A. Garfield went further, telling the House of Representatives that American scientists did not want the government to meddle in their affairs. "We have made the Government a formidable and crushing competitor of private students of science," he warned. "The Government ought to keep its hands off, and leave scientific experiment and inquiry to the free competition of those bright, intelligent young men whose genius leads them into the fields of research."[22]

Installed as director of the new, consolidated United States Geological Survey in 1879, Clarence King immediately forgot the hands-off policy, just as he ignored the statutory restriction that confined the survey to practical economic geology. He wanted a nationwide survey with a half-million-dollar budget. Scientists and politicians fearful of King's ambitious policy accused him of having imperialistic designs on state and local science. It was a "political wrong," an "infringement on State Rights," complained James Dwight Dana.[23] Right or wrong, however, by 1879 the states' rights

---

20. Cf. Thomas G. Manning, *Government in Science: The U.S. Geological Survey, 1867–1894* (Lexington, Ky., 1967), pp. 1–31.

21. U.S. Congress, House, *Survey of the Territories*, Miscellaneous Document, 45th Cong., 3d sess., Dec. 3, 1878, no. 5, p. 24; Manning, *Government in Science*, pp. 38–58.

22. U.S. Congress, House, *Congressional Record*, 45th Cong., 3d sess., February 11, 1878, p. 1209.

23. James D. Dana, "Geological Survey of the Public Domain," *American Journal of Science*, 3d ser. XVIII (November, 1879), 492–96; Clarence King, *First Annual Report of the United States Geological Survey* (Washington, D.C., 1880), pp. 3–6, 79; Thurman Wilkins, *Clarence King* (New York, 1958), pp. 230–38, 240–45.

argument was, in Roscoe Conkling's words, little more than a "faded sentimentality." When Congress finally turned against King, it was because of his budgetary demands, not because he wanted a centralized survey. Before he resigned in 1881, he secured the new president's pledge that John Wesley Powell would succeed him.[24]

By 1884 Powell had succeeded so well that President Cleveland's economy-minded administration decided it was once again necessary to rein in government science. Senator William B. Allison headed the congressional investigation, but Representative Hilary A. Herbert of Alabama was the driving force behind retrenchment. The former Confederate colonel bore no personal grudge against science or scientists. Rather, he was a man of principle, whose principles were out of joint with his times. "I believe in as little government as possible—that Government should keep its hands off," he explained to O. C. Marsh. "This is the doctrine I learned from Adam Smith & Mill & Buckle, from Jefferson, Benton and Calhoun, and from this stand point I believe we have to much to do (the Gov't) with pure science."[25] Herbert had been particularly impressed by Buckle's account of what had happened to science in the court of Louis XIV. Royal patronage, said the English historian, had stifled creativity and independent thought. Scientists had degenerated into mere courtiers, scrambling for royal favor.

Herbert found an ally—virtually his only one—in Alexander Agassiz. Launching an attack on the Geological Survey, they charged that it was a threat to laissez faire science. As the Allison Commission gathered testimony, Agassiz fed Herbert information and opinions, planted articles in *The Nation, Science,* and *Popular Science Monthly,* and appealed directly to the president. Anonymously, he told the readers of the *Popular Science Monthly* that "private enterprise, individual interest and effort, and voluntary association" were the safest stimulants to science. Privately, he condemned "the general and unlimited government coddling of science,"

---

24. U.S. Congress, Senate, *Congressional Record,* 46th Cong., 2d sess., May 10, 1880, pp. 3158–61; House, *Congressional Record,* 46th Cong., 1st sess., June 28, 1879, pp. 2420–24; Manning, *Government in Science,* pp. 60–68.

25. Herbert to O. C. Marsh, July 13, 1886, O. C. Marsh Papers, Peabody Museum, Yale University, New Haven, Conn.

and muttered about "federal pap liberally supplied to every crank who has a scheme."[26]

Agassiz's coaching allowed Congressman Herbert to act before other members of the Commission were ready. With a vigorous nod from the zoologist, he seized upon vertebrate paleontology as the least immediately useful and hence potentially most vulnerable phase of Survey activity. Appearing to speak for the entire Commission, in April, 1886, he reported out a bill which eliminated paleontological research entirely and curtailed Survey work as a whole. "If the Government is to claim the . . . field . . . , a monopoly will be established, private enterprise will be repressed, and competition will be destroyed."[27]

Herbert had actually spoken only for himself, for his Alabama colleague on the Commission, John T. Morgan, and for Alexander Agassiz. The Commission as a whole repudiated retrenchment. O. C. Marsh wrote to say that the scientific community opposed the measure, and regretted that he and Herbert had not had a frank conversation at the outset.[28] John Wesley Powell, now forced to defend his administration, contradicted his position of 1878 and argued that government science was both intrinsically good and a valuable stimulant to private investigations. "The laws of political economy do not belong to the economies of science and intellectual progress," he asserted, rejecting Herbert's Malthusian view of

---

26. [Alexander Agassiz], "Official Science at Washington," *Popular Science Monthly,* XXVII (October, 1885), 847; Alexander Agassiz to Thomas Huxley, April 17, 1886, Alexander Agassiz Papers, Museum of Comparative Zoology, Harvard University, Cambridge, Mass. (hereafter cited as MCZ); *idem,* "The Coast Survey and 'Political Scientists,'" *Science,* VI (September 18, 1885), 253–55; Agassiz to Hilary Herbert, November 23, 1885, MCZ; [Alexander Agassiz], "Science and the State," *Popular Science Monthly,* XXIX (July, 1886), 412–15; *idem,* "The National Government and Science," *The Nation,* XLI (December 24, 1885), 525–26; U.S. Congress, Senate, Joint Commission to Consider the Present Organizations of the Signal Service, Geological Survey, Coast and Geodetic Survey, and the Hydrographic Office of the Navy Department . . . , *Testimony,* Miscellaneous Document, 49th Cong., 1st sess., 1886, no. 82, serial 2345, pp. 1013–16 (hereafter cited as Allison Commission, *Testimony*).

27. U.S. Congress, House, *Restricting the Work and Publications of the Geological Survey, and Other Purposes,* House Report, 49th Cong., 1st sess., May 5, 1886, no. 2214, p. 12.

28. Marsh to Hilary Herbert, July 5, 1886, Marsh Papers, Yale; Abram S. Hewitt to O. C. Marsh, May 3, 1886, Marsh Papers, Yale; Simon Newcomb to Alexander Agassiz, June 26, 1886, MCZ.

knowledge. "The learning of one man does not subtract from the learning of another, as if there were a limited supply of unknown truth. Scholarship breeds scholarship, wisdom breeds wisdom, discovery breeds discovery." As for Agassiz, Powell dismissed him as a jealous aristocrat, frustrated because he had been unable to dictate the course of American science from the Agassiz museum at Harvard.[29]

Powell's criticism of Alexander Agassiz was unfair. Agassiz's motives were more complex, a mélange of hypocrisy, envy, and political naïveté. Agassiz believed in self-made manhood. He had deliberately set about making a fortune in Michigan copper so that he could, while still a young man, retire from business to a life in science. He financed his own expeditions, published his own researches, paid his own way, and never quite understood why others did not follow his example. Agassiz modestly discounted his own extraordinary talent. He also forgot that growing up in his father's household, attending the Lawrence Scientific School, and inheriting a position in the upper reaches of Boston society all gave him something of a favored position in the struggle for professional life.[30]

Agassiz was envious, but not of government science. Like his father, he had benefited from a close liaison with the Coast Survey, and had carried out extensive investigations under a mutually advantageous cost-sharing plan. When the Coast Survey was involved, Agassiz was willing to urge liberal appropriations for government science.[31] The real object of his envy was O. C. Marsh, who had gained a national reputation as a patron of science as well as an investigator. Agassiz, on the other hand, had not. Marsh merely spent his uncle George Peabody's money, while Agassiz made his own. As official paleontologist of the Geological Survey, Marsh had mingled public and private funds and enhanced his Yale collections

---

29. Allison Commission, *Testimony*, pp. 1070–78; Dupree, *Science in the Federal Government*, pp. 224–31.

30. George R. Agassiz, *Letters and Recollections of Alexander Agassiz, with a Sketch of His Life and Works* (Boston, 1913), demonstrates the need for a full modern biography.

31. Alexander Agassiz, "The Coast Survey and 'Political Scientists,'" p. 255.

at government expense. He was, in short, "passing himself upon the world as a great benefactor of paleontology," when in reality he was spending other people's money. Agassiz's bitter opposition to government science—that is, to certain kinds of government science—was more a function of hurt feelings than of political philosophy.[32]

In the end, Alexander Agassiz and Hilary Herbert were out-politicked by "political scientists" in Washington and outflanked by the social and economic tendencies of their times. In its final report, the Allison Commission did little more than request that John Wesley Powell tidy up his accounting procedures. Herbert, the politician, took the defeat in stride. Agassiz, the scientist-turned-political economist, retired in disgust to the calm of oceanographic research. Except for a brief sally in 1892, when Marsh and government paleontology were once again under fire, he never entered the political arena again. The Geological Survey, and everything it represented, was here to stay.[33]

The absence of significant conflict between public and private scientific enterprise in Gilded-Age America stood in revealing contrast to the controversy in Victorian England, the one major European nation whose scientific institutions had a roughly similar fiscal structure. Ever since the 1830s, when Charles Babbage and others had blamed the tradition-bound Royal Society for the poor showing of British science, the emerging scientific community had agitated for increased research subsidies.[34] The British Association for the Advancement of Science (1831) led the campaign, enlisting such unlikely allies as Herbert Spencer and Prince Albert. In 1859, Albert told the association that they should face matters realistically,

---

32. Agassiz to H. N. Mosley, September 30, 1885, MCZ; Agassiz to Thomas Huxley, October 14, 1885, MCZ.

33. Agassiz to George Brown Goode, February 5, 1888, MCZ; Agassiz to Edward O. Wolcott, July 2, and July 20, 1892, MCZ; Manning, *Government in Science*, pp. 122–50, 204–16.

34. Charles Babbage, *Reflexions on the Decline of Science in England* (London, 1830), *passim*; [Sir David Brewster], "Review of Babbage's *Reflexions on the Decline of Science in England*," *The Quarterly Review*, XLIII (October, 1830), 305–42; Nathan Reingold, "Babbage and Moll on the State of Science in Great Britain: A Note on a Document," *British Journal of the History of Science*, IV (June, 1968), 58–64. Donald S. L. Cardwell, *The Organisation of Science in England* (London, 1957), is a useful survey.

stop feeling sorry for themselves, and "hand round the begging-box and expose themselves to refusals and rebuffs . . . with the certainty besides, of being considered great bores." [35]

The English debate over public versus private science began in earnest late in 1868, when Alexander Strange, an engineer and fellow of the Royal Society, proposed outright state subvention of research. Strange called for a royal commission to investigate the entire problem, in order to determine whether it was in fact true that public subsidies would "chill private enterprise." [36] The scientific community itself was divided. Henry Roscoe, one of England's leading chemists, said that Germany was argument enough for government science. Alfred Russell Wallace disagreed, claiming that state aid was "radically vicious" in principle and disastrous in practice. Thomas Henry Huxley supported reform by criticizing the status quo. "If continental bureaucracy and centralisation be fraught with multitudinous evils," he observed, "surely English beadleocracy and parochial obstruction are not altogether lovely." Alive to the mounting controversy, in 1872 the Crown appointed the Devonshire Commission. [37]

The Commission gathered testimony for three years, and even interviewed the vacationing Joseph Henry, who added little to the discussion. The Commission's activities kept the issue alive in the press. Liberal politicians tended to support laissez faire science on the grounds that it embodied the British principle of a fair field and no favors. Most scientists, however, seemed eager to accept a few favors. When Prime Minister Gladstone and the *Times* opposed

---

35. *Report of the British Association for the Advancement of Science, 1859* (London, 1860), pp. lxvii–lxviii.

36. Alexander Strange, "On the Necessity for State Intervention to Secure the Progress of Physical Science," *Report of the British Association for the Advancement of Science, 1868* (London, 1869), pp. 6–7; *idem*, "On the Necessity for a Permanent Commission on State Scientific Questions," *Nature*, IV (June 15, 1871), 130–33.

37. Henry E. Roscoe, "Scientific Education in Germany," *Nature*, I (December 9, 1869), 157; Alfred Russell Wallace, "Government Aid to Science," *Nature*, I (January 13, 1870), 288–89, and (January 20, 1870), 315; Thomas H. Huxley, "Administrative Nihilism," *Nature*, IV (October 19, 1871), 495–96; *Nature*, IV (October 12, 1871), 461–62.

public subsidies for science, the editor of *Nature* condemned such "mental darkness in high places." [38] In its final report the Commission declared that the days of Faraday and Davy were gone, as were the days of virtuosi and gentleman naturalists. Science had become a national resource, a foundation of the general welfare. As such it had, and would increasingly have, a claim upon both the private and the public sectors. [39]

In England as in the United States, public and private science had developed in pragmatic partnership. The principal difference in the two nations' experience was that when the question of inherent conflict was raised, the Devonshire Commission seriously examined the larger questions of public policy involved. By contrast, the Allison Commission (except for the quaintly Jeffersonian Hilary Herbert) seemed content merely to tinker with the administrative machinery of government science, and then only in the name of economy.

The Allison Commission had reaffirmed a long tradition. As William B. Clark would remind the American Society of Naturalists in 1900, the political economy of American science had always turned on matters of expediency, not abstract principle. More than they knew, and probably more than they would have cared to admit, professional men of science shared certain working principles with nineteenth-century bread-and-butter unionism. They worked for concrete goals, which by advancing knowledge also advanced their own careers. As Adolph Strasser of the cigar-makers' union phrased it: "We have no ultimate ends. We are going on from day to day. . . . We want to dress better and to live better. . . . We are all

---

38. *Times* (London), April 26, 1872; *Nature*, VI (May 2, 1872), 1–2 (May 9, 1872), 21–22; "The Endowment of Research," *Nature*, VIII (June–September, 1873), *passim;* George Gore, "The National Importance of Scientific Research," *Westminster Review,* XCIX (April, 1873), 163–73; C. E. Appleton, ed., *The Endowment of Research* (London, 1876), *passim;* Donald S. L. Cardwell, "The Development of Scientific Research in Modern Universities: A Comparative Study of Motives and Opportunities," in *Scientific Change,* ed. A. C. Crombie (New York, 1963), pp. 661–77.

39. Royal Commission [The Devonshire Commission] on Scientific Instruction and Advancement of Science, *Final Report* (London, 1875), summarized in *Nature,* XII (August 12, 1875), 285–88.

practical men." [40] American men of science were also practical men. They understood that the most significant developments in the science of their day—the increase of technical knowledge, the process of professionalization, the founding of institutions, and the creation of a reliable fiscal structure—were all matters which transcended or ignored the conflict between public and private scientific enterprise.

40. "The Attitude of the State toward Scientific Investigation," *Science,* n.s. XIII (January 18, 1901), 81–95; John R. Commons et al., *History of Labour in the United States,* 2 vols. (New York, 1926), II, 309.

II

Nineteenth-Century American Scientists and Their Research

Was There a Scientific Lazzaroni?

An experienced hand at the politics of science was T. Romeyn Beck of Albany. He was also an old and trusted mentor of Joseph Henry. His letter to Henry about the early work of the Smithsonian Institution reflected both his experience and their comradeship. Beck directed Henry's attention to a danger which "may render your scheme unpopular, and hence in a measure impair its usefulness. It is the possibility of the . . . formation of predominant cliques. These are the curse of most of our distinguished societies at home and abroad—and in this country the danger is great, from the fewness of men well grounded in science, and the disparity that exists between those claiming to be adepts." [1] Beck's warning stemmed from his knowledge of society and of the particular temptations surrounding the launching of a new scholarly venture. He wisely sensed that Henry, regardless of his efforts to the contrary, might be accused of fostering a clique. Even if a clique did not form, Henry's success in preventing one might go unnoticed, for few traits seem more American than the ability to see a cabal where none exists. Our propensity to see human intelligence underlying turmoil and guiding change seems particularly strong during periods of severe social unrest. When the present is frightening and the future uncertain, we identify the sources of our social strife in the

---

1. T. Romeyn Beck to Joseph Henry, November 29, 1847, reproduced in William J. Rhees, "The Smithsonian Institution: Documents Relative to Its Origin and History," *Smithsonian Miscellaneous Collection*, XVII (1880), 961.

avarice and hate of other men. Few periods of our social history have been more conducive to imagining conspiracies than the years of our Civil War and the decade preceding it. Civil discord frequently made civil discourse impossible, and there was ample evidence that men were indeed conspiring against one another.

In such an era it is not surprising that the political and social mood should infect relationships among men who considered themselves immune to rancor. Men of science were becoming aware that science, like slavery, had a politics, and that the politics of science might be as brutal as the politics of abolition. The Smithsonian Institution and the American Association for the Advancement of Science had been recently founded. The Coast Survey and the Naval Observatory were flourishing. Silliman's prestigious *American Journal of Science* had recently relaxed its editorial control to include younger men; and there were encouraging signs that education in science might become firmly rooted in Cambridge, New Haven, New York, Albany, Philadelphia, and Ann Arbor. The men who controlled those institutions and agencies would profoundly influence both the quality and quantity of science in this country. Put simply, there was, during the 1850s, a rather sudden and extensive increase in attractive scientific prizes.

It is therefore understandable that in those times some aspiring savants ascribed their professional frustrations to organized plots against them. John Warner, for example, corresponded extensively with persons who sympathized with his claim that a "Cambridge-Washington clique" sought "scientific dominion" by fair means and foul. The correspondents of this Philadelphia physician agreed that "the scientific aristocrats and snobs at Cambridge . . . have mistaken themselves for Gods." They saw the Smithsonian Institution as "a central establishment for the control of science." Some wanted to fight the clique by rebelling at AAAS meetings. Others supported Matthew Fontaine Maury, who simply refused to fight, dropped his membership, and continued his investigations.[2]

---

2. John Warner Papers, American Philosophical Society, Philadelphia, Pa. (hereafter cited as APS). See especially C. F. Winslow to Warner, September 3, October 2, October 24, and November 21, 1857, and February 23, 1858; D. A. Wells to Warner, June 25, 1859.

Better-known examples of this kind of criticism, leveled at essentially the same group of men by scientists who were their peers, may be found in initial reactions to the establishment of the National Academy of Sciences. Louis Agassiz, Alexander Bache, and others had dreamed aloud about such an organization for almost fifteen years prior to its founding. They had in mind an agency that would allow scientific issues to be settled among scientists rather than politicians. In addition, they sought to raise the unified voice of American science when public policy required exact knowledge. An organization with such goals was an obvious target for abuse and suspicion by those not privy to its inner councils.[3]

Scientists and politicians who feared a clique of ambitious entrepreneurs of science were not paranoid. There was ample evidence to fuel dark thoughts. The same men could be seen, year after year, nominating, seconding, and electing succeeding presidents of AAAS. At the gala social affairs surrounding AAAS meetings, they, with their friends, protégés, and consorts, would dine with the business elite of the host city. It was common knowledge that they were well connected in their home towns—that Agassiz and Benjamin Peirce enjoyed membership in the Saturday Club along with Holmes, Longfellow, and Emerson; that Bache and Henry were on easy terms with such powerful politicians as Jefferson Davis, Henry Wilson, and Charles Sumner; that all were welcome at the Wistar parties in Philadelphia and among the powerful friends of George Templeton Strong and Samuel Ruggles in New York City. What was not so well known was that this clique had a name. It had, in fact, two names. To its friends, such as Strong, Ruggles, and Francis Lieber, it was the Florentine Academy. To its members it was the Lazzaroni.[4]

---

3. See, for example, J. P. Lesley to B. S. Lyman, May 7, 1863, Lesley Papers, APS; Lesley to his wife, April 23, 1863, in *Life and Letters of Peter and Susan Lesley*, ed. M. L. Ames (New York, 1909), p. 419; and W. B. Rogers to Henry Rogers, April 28, 1863, in *Life and Letters of William Barton Rogers*, ed. Emma Rogers and W. Sedgwick (Boston, 1896), p. 161.

4. The Spanish word *lazarino*, meaning *leper*, entered Italian as *lazzaro* in the early seventeenth century. In Italian the word referred originally to the lower-class populace of Naples, the "plebians of the marketplace." Gradually it came to refer to any persons of humble origin or low caste who were restless or in revolt. These *lazzaroni* developed their own folk heroes, leaders legendary for their ability to in-

These Lazzaroni have proven as intriguing to recent historians as they did to scientists who were contemporaries. A symbol of the mid-nineteenth-century effort to professionalize and specialize the pursuit of scientific knowledge, the Lazzaroni as a group have been vested with a greater coherence and energy than in fact existed. Because several members were indeed eminent scientists by standards of their day and ours, the group itself has taken on a scholarly luster that exceeds its collective merit. This chapter will examine the Lazzaroni as they were and discuss to what extent the group followed its self-proclaimed objectives. What the Lazzaroni were will be contrasted to what they were not. Only in this way can their reality be compared to the life that historians have breathed into them.[5]

WHO WERE THE LAZZARONI?

The first use of the name *Lazzaroni* seems to be in a letter from Alexander Dallas Bache to Cornelius Conway Felton just before Christmas of 1850. Bache already referred to himself as "the Chief," and listed seven persons in addition to himself and Felton. Subsequent letters from two other members acknowledged those nine names. All agreed that the group consisted of Alexander Dallas Bache, Joseph Henry, John Fries Frazer, Wolcott Gibbs, Benjamin Apthorp Gould, James Dwight Dana, Louis Agassiz, Benjamin

---

spire and organize the downtrodden masses. However, like their scientific namesakes in the United States, the Italian lazzaroni were not always found fighting in the cause of liberalism and popular rights. At times, as in the late eighteenth century, they cooperated with factions of Italian nobility in repressing the ambitious middle classes. After the victory of the Swiss Guard in Italy in 1848, the lazzaroni willingly shared in the spoils.

5. The most significant discussion of the Lazzaroni in the recent literature is Edward Lurie, *Louis Agassiz: A Life in Science* (Chicago, 1960), pp. 181–84. See also A. Hunter Dupree, *Science in the Federal Government* (Cambridge, Mass., 1957); and *idem, Asa Gray* (Cambridge, Mass., 1959). Lurie's conclusions about the group are reflected in Nathan Reingold, *Science in Nineteenth Century America* (New York, 1964); and Howard S. Miller, *Dollars for Research* (Seattle, Wash., 1970). The following biographies of scientists also include assertions that the clique existed and that it sought to control American science: Frances L. Williams, *Matthew Fontaine Maury, Scientist of the Sea* (New Brunswick, N.J., 1963), p. 236; Merle Odgers, *Alexander Dallas Bache, Scientist and Educator, 1806–1867* (Philadelphia, Pa., 1947), p. 204; and Thomas Coulson, *Joseph Henry* (Princeton, N.J., 1950), II, 157.

Peirce, and Cornelius Conway Felton.[6] Bache, Frazer, and Peirce wrote to one another frequently and at length. Bache, Henry, Agassiz, and Gibbs traveled extensively and consulted with one another en route to and at conferences. Frazer saw more of each man than any other member because his home in Philadelphia was a convenient stopover between Washington and New England. Dana, on the other hand, rarely wrote, seldom traveled, and apparently did not consider himself a part of the group. He and Gibbs saw each other occasionally in connection with their editorial duties for the *American Journal of Science,* and were included largely because of the power represented by access to that publication.

The Lazzaroni were men who lived comfortably, dined sumptuously, and drank heartily. Their avowed *raison d'être* was to meet annually, usually in January, to feast upon oysters and dark lager. That was perfectly in keeping with their individual personalities. James Russell Lowell remembered Felton "shaking in burly mirth" at meetings of the Saturday Club. Agassiz enjoyed an unparalleled reputation for repartee, and Gould was introduced even as a young man as "a glittering specimen of youthful America . . . who has consumed gallons of midnight oil." Many of their letters to one another speak as much of tender oysters and foolproof home brew as they do of science. Most of them delighted in calling one another by pet names. Everyone was serious in calling Bache "the Chief," but quite jovial in dubbing Henry, Peirce, and Agassiz, "Smithson," "Functionary," and "Fossilary," respectively. Bache and Frazer were mentor and student, and thus were "Grandpa" and "Grandson" in friendly discourse.[7]

Many of the Lazzaroni were affiliated with one another by circumstances other than mutual congeniality. Virtually all of them had some research supported by Coast Survey funds allocated by Bache

---

6. Bache to Felton, December 19, 1850, Bache Papers, Smithsonian Institution Archives, Washington, D.C.; Peirce to Frazer, November 2, 1858, and Gould to Frazer, November 22, 1856, in Frazer Papers, APS; Gould to Dana, November 22, 1856, in Dana Papers, Beinicke Library, Yale University, New Haven, Conn.

7. Edward W. Emerson, *The Early Years of the Saturday Club* (Boston, 1918); George S. Hilgard and Cornelius Conway Felton, *Proceedings of the Massachusetts Historical Society,* X (1869), 352–68; Peirce to Bache, October 28, 1853, Rhees Collection, Huntington Library, Pasadena, Calif.

and published their results in space provided by Henry. Gould was usually on the Coast Survey payroll, as were numerous other bright, well-connected young men whose personalities foreclosed faculty positions. Further, Peirce, Bache, Agassiz, and Felton were at one time or another regents of the Smithsonian. Their duties in that role kept them in intimate contact with influential persons in Washington and allowed them to lend their prestige to Henry's decisions about staff changes and publishing policies. Bache and Henry were close personally as well as professionally. They were the oldest of the nine, and their friendship dated to the early 1840s, when Henry was at Princeton and Bache was in Philadelphia. The three Harvard Lazzaroni were just as close personally. Felton, Peirce, and Agassiz were all neighbors in Cambridge. Felton and Agassiz were married to sisters, as were Peirce and the Lazzaroni fellow traveler, Charles Henry Davis.

Most of these nine men might have sought one another's company even if they had no common goals. They attached little significance to their membership in the group. As men of stature they needed no additional affiliation to enhance their prestige or insure themselves professional opportunities. Only Gould, the youngest and most abrasive, obviously needed the others as patrons. Significantly, Gould was the most aggressive member in trying to organize dinners and in reminding the others of their loyalties. Gibbs, Henry, and Agassiz were tolerantly bemused by the whole affair, while Peirce and Frazer found it a useful ornament in their correspondence with the other seven. Dana and Felton apparently took no interest. Each was included *ex officio* by the others for professional as well as personal reasons: Dana, because he was editor of the *American Journal of Science,* and Felton, because he was heir apparent to the Harvard presidency.

Was the Lazzaroni merely an expression of personal needs for congenial colleagues? Or did it represent a concerted, carefully planned effort to accomplish specific objectives with respect to American science? How does a historian ascertain the nature of a group? One way is to seek the ties of friendship and kinship which bind its members. Another is to inquire about the social and professional context in which its members worked. For instance, what opportunities were there in the 1850s for any group of willful men to control American science? What goals would such a band have had, and how would its members have conspired to attain them?

To dominate American science in the 1850s, any thoughtful group probably would have tried to do most of the following: first, control federal patronage by securing key appointments in the Smithsonian, Coast Survey, Naval Observatory, Patent Office, military academies, and similar agencies; second, control the primary outlets of scientific writing such as the *American Journal of Science,* AAAS proceedings, Coast Survey and Smithsonian publications, and publications of significant local institutes, academies, and surveys; third, write and speak extensively, with a clear, coherent, and consistent message, for both professional and popular audiences; fourth, try to place men in whom they had confidence in the elective offices of significant organizations, in key college professorships, and on important state surveys and scientific commissions; and, finally, be instrumental, or at least influential, in founding new scientific organizations or in reconstituting old ones. These opportunities might be seen as the potential sphere of influence. If intrigue was in the air, it should be apparent at many points within that sphere, even to the historian left with an imperfect record. Let us place the nine Lazzaroni in that context to examine their congruity and purpose as a group.

THE LAZZARONI AND EDUCATION

Many of the possible activities and interests mentioned above were related to education, and particularly to higher education. True politicians of science, like those of other domains, would have seized virtually any opportunity to couple their goals with American faith in schooling. By the 1850s that faith had grown so strong that references to the virtues of schooling were required in every politician's repertoire of public phrases. The Lazzaroni shared in that culture, and the group showed signs early in its existence of accepting the norms concerning education. But the group's rhetoric respecting education deteriorated rapidly to mindless jargon, and they took no effective action respecting education.

Consider, for example, the involvement of some of the Lazzaroni in the ambitious effort to establish a national university in Albany, New York. James Hall had been trying for several years to excite local interest in the scheme. But it was not until Agassiz consented to attend an organizational meeting in Albany in January, 1851, that there were some prospects of success. On the basis of that meeting and the knowledge that the AAAS would meet in Albany that summer, the local newspaper proudly announced plans for the new university. The state legislature obligingly chartered the institution in April.[8]

By midsummer, the *Daily Albany Argus* was describing the venture to an admiring public and asserting that the new institution would rival any in Europe because it would foster "professional and profound research in all departments of human knowledge." More significant, Agassiz was assuring Hall that "my whole soul is bent upon the project of an American university" and that there was "no reasonable sacrifice I am not ready to make for its establishment." The friendship between Agassiz and Hall was the chief link between the local boosters and the larger world of science. Word was out that Agassiz might indeed leave Cambridge for Albany. The hope was that Agassiz would attract other scientists to the enterprise, especially Professors Peirce, Henry, and Bache.[9]

It was Bache who set the tone for the discussion of the project among his colleagues. In his presidential address to the AAAS, he called for more professionalism and specialization in science. These goals, Bache argued, would be attained only when scientists became better organized and when there was established "an institute of science . . . to guide public action in reference to scientific matters." Thus Bache publicly began his campaign to secure what twelve years later became the National Academy of Sciences. But he also specifically disclaimed any relationship between institutions founded to teach science and the advancement of science through

---

8. *Daily Albany Argus*, February 6, 17, and 22, 1851; Circular of the Scientific Department of The University of Albany (Albany, N.Y., 1851).

9. *Daily Albany Argus*, June 4 and 17, July 26 and 29, August 1 and 4, 1851; Agassiz to Hall, August 3, 1851, quoted in John M. Clarke, *James Hall of Albany* (Albany, N.Y., 1828), p. 193; Spencer Baird to Agassiz, July 14, 1851, in *Correspondence between Spencer Fullerton Baird and Louis Agassiz*, ed. Elmer Herber (Washington, D.C., 1963).

basic research. In so doing, he revealed a fundamental disagreement of which the Lazzaroni themselves were hardly aware. The present generation of academics knows this disagreement as the conflict between research and teaching. The Lazzaroni and their contemporaries eventually came to understand it in much the same way.[10]

Benjamin Peirce, for example, affirmed his commitment to the Albany university and noted it must be "in strict harmony with the wants of the country." For Peirce, those wants were best illustrated by the burgeoning system of common schools. The university must be the capstone of that system, a place "equally accessible to all students of every condition of life." Peirce's list of persons whom he considered essential for the faculty included such men as John Pitkin Norton, who could teach agriculture to the "practical farmer." Indeed, it was Norton's series of lectures that represented the official opening of the institution.[11]

Another index of opinion on the teaching-versus-research issue was willingness to adhere to the tempting fee system for faculty remuneration. The notion that professors should be paid directly by their students, thereby equating income with popularity, was taken seriously in this country after it was endorsed by Francis Wayland in his influential *Thoughts on the Present Collegiate System,* published in 1842. Peirce was quite willing to begin under that arrangement, while the venerable Agassiz thought it should apply only to the young professors. Dana, on the other hand, could conclude that "a grand university" was in the offing only after being convinced that the state would support it by funding students as well as paying professors.[12]

---

10. Alexander Dallas Bache, Presidential Address, *Proceedings of the American Association for the Advancement of Science,* VI (1851), 41–60.

11. Peirce to Hall, October 19, 1851, Hall Papers, Manuscript Division, New York State Library, Albany, N.Y.; on Norton's activity in Albany, see George Brush to William Brewer, October 21, 1851, Brush Papers, Sterling Library, Yale University, New Haven, Conn.; and *The Cultivator,* VIII (1851), 344, and IX (1852), 64.

12. Francis Wayland, *Thoughts on the Present Collegiate System in the United States* (Boston, 1842); Peirce to Hall, October 19, 1851, Hall Papers; Agassiz to Dana, February 9, 1852, Agassiz Papers, Houghton Library, Harvard University, Cambridge, Mass.; Dana to John L. LeConte, February 21, 1852, LeConte Papers, APS. For a more complete description of the Albany university venture, see Robert Silverman and Mark Beach, "A National University for Upstate New York," *American Quarterly,* XXII (1970), 701–13.

There was no unified enthusiasm among the Lazzaroni for the Albany enterprise. The prospects in Albany both appeared and faded early in the decade. Even among those who were interested there was no consistent point of view on the nature of the institution. An examination of later public statements about the idea of a true university made by four of the group in 1856, four years after the Albany affair and in a period of great agitation for the reform of higher education, shows their lack of consensus. In speeches and publications by Dana, Peirce, Gould, and Bache, nothing remotely resembling a party line emerged beyond common allegiance to something called "a university." Dana would have only permanent professors with endowed salaries and would teach metallurgy and veterinary medicine, as well as "science for its own sake." Peirce, on the other hand, stressed migrating professors who would be required to give at least twelve lectures to be considered members of any one faculty. Benjamin Gould did not discuss the possibility of an itinerant faculty, but stressed that they must be exclusively concerned with developing researchers rather than engineers. Finally, Bache contended that the faculty could teach only through the example set by its exemplary research. The professors might never lecture or even be physically present.[13]

Another means of searching for a Lazzaroni position on education is to sample the thoughts of one member over a fairly long time period. One would expect those thoughts to be reasonably in tune with theorizing by leading educators as well as to be consistent with the statements of other Lazzaroni. Consider the statements of Joseph Henry from 1852 (when the group was reputedly well established) through 1873 (when the group was widely scattered). As early as 1852, Henry had noted that he and Frazer strongly disagreed over

---

13. James Dwight Dana, *Proposed Plan for a Complete Organization of the School of Science connected with Yale College* (New Haven, Conn., 1856); Benjamin Peirce, *Working Plan for the Foundation of a University* (1856), privately printed and marked "confidential"; Benjamin A. Gould, *An American University* (Hartford, Conn., 1856); Alexander Dallas Bache, "Remarks at the Opening of the Fifth Session of American Association for the Advancement of Science," *American Journal of Education*, I (1856), 477–79; *idem, Address before the American Institute of the City of New York, October 28, 1856* (New York, 1857).

the teaching-versus-research issue. Frazer contended that researchers would dwell at length on their narrow concerns, would not keep in touch with wider scholarly interests, and would thereby be ineffective teachers. Henry feared that opinion might prevail as efforts were made to reorganize the University of Pennsylvania, so he urged Bache to use his influence in Philadelphia on behalf of true science.

In addition to disagreement within the group on educational issues, there is ample evidence that Henry, as well as other Lazzaroni, was ignorant of significant portions of contemporary thought about education. For example, in 1854, Henry had been conscripted as president of the newly founded American Association for the Advancement of Education. He began his presidential address by pointing to his own ignorance of the field of education, then proceeded to demonstrate the truth of that allegation against himself. Henry insisted on the value of constant drill at all levels of schooling as the only way to train the faculties. While some other Lazzaroni were calling for institutions that would free youth to learn practical arts, Henry sought to tighten the intellectual strait jacket in which students were confined. He stressed the importance of forming good habits and the difficulty of diffusing knowledge in light of man's natural indolence. Despite criticism from even the conservative organizers of the conference during three hours of discussion following his speech, Henry republished his remarks a year later virtually unchanged.[14]

It is tempting to think that Henry's strident advice was intended for schools instead of universities. Clearly that was not the case. Henry's counsel to persons in higher education stressed the same themes. Students should be lectured to, drilled, and tested. Only in his views of the relationship between teaching and research did he become somewhat more flexible over the years. By 1873, he was willing to concede that teaching duties might force a man to keep in

---

14. Henry to Bache, July 8, 1852, Henry Papers, Smithsonian Institution Archives; Henry, "Thoughts on Education," *American Journal of Education*, I (1856), 17–31; *idem*, entry for December 29, 1854, locked diary in Smithsonian Institution Archives; and *idem*, "Thoughts On Education," *The Spectator*, December 5, 1855.

mind the fundamental facts and principles of his field. Thus "even the most profound investigator . . . in imparting information to others would gain clearer conceptions himself."[15]

There is one more way to probe for a coherent Lazzaroni view of education: examination of the opportunities for influence open to the group in a decade greatly concerned with colleges and universities. The prestigious eastern colleges were not the only institutions that showed promise of becoming true universities. In farsighted states such as Virginia and Michigan, there were ambitious scholars striving to re-create the research-oriented institutions they had encountered in Germany. The Lazzaroni were silent about these efforts in their letters to one another. They were also silent about the repeated attempts to secure congressional land grants in support of education in scientific agriculture. There is no evidence that they noticed the first Morrill Act when it finally passed, much less that they tried to influence its legislative fate during the four years between its proposal and passage. Yet men such as Bache and Henry had well-deserved reputations for political awareness and effective relations with Congress. More significant, the Lazzaroni hardly expressed themselves on pressing issues of education even when raised by their friends and colleagues. For example, Samuel Ruggles' stately analysis of the duty of Columbia College to the community went unnoticed, although it was a direct outgrowth of Ruggles' fight on behalf of a professorship for Wolcott Gibbs. The lucid pronouncements of Francis Lieber and John Draper went equally ignored. While some Lazzaroni consented occasionally to speak on the topic of education, others rarely found it worth the effort.[16]

---

15. Henry, "Notes on the Plan of Organization of the University of the South," clipped to letter to Bache, January 9, 1860, Bache Papers, Smithsonian Institution Archives; *idem,* "An Examination of Professor Henry by the English Government Scientific Commission," *Smithsonian Institution Miscellaneous Collection,* XVIII (1880), 775–801; *idem,* "Address at the Opening of a School of Science at Princeton," manuscript in Henry Papers.

16. Samuel B. Ruggles, *The Duty of Columbia College to the Community, and its Right to Exclude Unitarians from its Professorships of Physical Science* (New York, 1854); *Testimonials presented to the Trustees of Columbia College in behalf of Dr. Wolcott Gibbs . . . January 9, 1854* (New York, 1854); Milton H. Thomas, "The Gibbs Affair at Columbia in 1854" (Master's essay, Columbia University, 1942).

One has only to examine the efforts to establish a university in New York City in 1856 to realize the indifference of most Lazzaroni to proposals dealing with education. Peter Cooper stood ready to put up a munificent sum for an institute of science. Here was a golden opportunity to create a true university by uniting Cooper's establishment with the Astor Library, the Free Academy, and Columbia College. The idea had the blessing of Mayor Fernando Wood and such scions of New York society as George Templeton Strong. Moreover, Henry Tappan was willing to return from Ann Arbor to translate the grand scheme into action. If the support of the scientific community could be mobilized, success would be assured. Tappan wrote repeatedly to the leading Lazzaroni imploring them to assist. Lieber insisted to Bache that the advancement of science required support of universities. Urged by Lieber to support the philanthropist's plans, Gould spent a long day on a train with Cooper. Gibbs in New York City stood to gain immensely from the proposed arrangement. Yet the Lazzaroni response was conspicuous silence. Aside from a few halfhearted encouragements from Bache, they seemed not the least interested in the whole affair. Their lack of interest explains why no member of the Lazzaroni or their immediate circle was mentioned in Peter Cooper's autobiographical account of the founding of the Cooper Union.[17]

Some Lazzaroni were, of course, concerned with education when it involved securing proper jobs for men who met with their approval. Most were happy to testify on behalf of Gibbs to the Columbia trustees. And the installment of Gibbs in the Rumford chair at Harvard in 1863 represented the zenith of Lazzaroni power in

---

17. Henry Philip Tappan to Bache, June 19, 1856, and Gould to Bache, February 11, 1857, Bache Papers, Smithsonian Institution Archives; Francis Lieber to Samuel Ruggles, April 4, 1856, and through their many letters during the next two years, Lieber Papers, Library of Congress, Washington, D.C.; Tappan to Peirce, April 5, 1856, Peirce Papers, Harvard Observatory, Cambridge, Mass.; Tappan to Ruggles, January 29, February 19 and 20, and March 3, 1956, and Tappan to William Astor, February 18, 1856, all in University of Michigan Archives, University of Michigan, Ann Arbor, Mich.; Allen Nevins and Milton Thomas, eds., *The Diary of George Templeton Strong* (New York, 1952), II, 339. See also Alexander Dallas Bache, *Address before the American Institute of the City of New York, October 28, 1856* (New York, 1857); Peter Cooper, "Autobiography of Peter Cooper," typescript (1882), Cooper Union Library, New York, N.Y.; and Thomas Hughes, *Life and Times of Peter Cooper* (London, 1886).

Cambridge. Agassiz and Peirce were then able to accomplish for Gibbs what they had failed to bring about for Gould in 1859, when the directorship of the Harvard Observatory had been vacant. But even in the matter of placing men in jobs, there is no evidence of conspiracy. Neither Henry nor Bache, for example, hesitated to rank an outsider above one of their own. At least for these two members, Lazzaroni affiliation was not a primary job qualification.

Lazzaroni involvement in education is significant for two reasons. First, in the minds of most leaders in both fields, the advancement of education was increasingly linked to the growth of science. Men seeking to function as an elite of science ought to seize every opportunity to turn a venture in education to their own purposes. Instead, Henry's admission, late in his career, that he had only the vaguest knowledge of the origins and progress of the land-grant universities revealed a characteristic Lazzaroni indifference to education. Second, many members of the Lazzaroni were under constant pressure to accept professorships themselves, to recommend others for those posts, and generally to mobilize American science behind one university scheme or another. They responded to this pressure by either ignoring it or taking the minimum necessary action.

POLITICS OF SCIENCE: AAAS

There are other issues and events which reveal the extent, if any, to which the Lazzaroni had any rational or coherent program. Let us look briefly at three: (1) control of the AAAS, (2) the establishment of the Dudley Observatory, and (3) the founding of the National Academy of Sciences. Each offered a different opportunity for influencing American science. Each engendered controversy among scientists. Each involved some members of the Lazzaroni, but none involved all.

The leading Lazzaroni were concerned from the outset that AAAS leadership must not include "pretenders to science." The presidency was especially prestigious, so it is not surprising that they sought and gained control of that office in the organization's formative years. For eight years after W. C. Redfield chaired the organizing meeting in 1848, a member of the Lazzaroni or a man whom they approved was president. Henry, Bache, Agassiz, Peirce, and Dana all served as AAAS president. The Lazzaroni and their protégés also gave a large share of the papers and filled the majority of the committee positions during those early years. Some were especially careful to seek influence over anything the AAAS would publish. Nevertheless, as the organization grew larger and more complex, some Lazzaroni felt ready to abandon efforts to control or even support it. Gould recorded the group joke that AAAS signified "Amazing Asses Adverse to Science" and wrote that "more could be done with one good feed with decent fellows." But then, Gould was young, abrasive, and hard to please.[18]

In fact, Gould was at the center of the most vitriolic controversy in which some Lazzaroni engaged with outsiders. Indeed, the vexing struggle over the Dudley Observatory in Albany was the first time a part of the group had been formally involved in defending its point of view. Further, the enemy were not the scientists of dubious merit or petty politicians who had previously frustrated individual Lazzaroni. This time they were men of means with the cunning and experience to carry out their threat to bring the fight to the floor of the United States Congress. If ever there was a time for all nine men to rally, it was then.

POLITICS OF SCIENCE: THE DUDLEY OBSERVATORY

In 1851, land in Albany was donated for an observatory for the university. One donor of the land, along with other leading citizens, later gave funds for a building and some apparatus. Two governing bodies were constituted: a board of trustees, comprised of prominent local businessmen, and a scientific council, made up of Bache, Henry, and Peirce. All parties clearly understood that the observatory was for scientific, commercial, and educational purposes. Instruments were purchased and rules were established to serve those three goals.

---

18. Agassiz to Frazer, August 7, 1851, and July 16, 1852, Frazer Papers, APS; Peirce to Bache, August 30, 1853, Peirce Papers; Agassiz to Bache, August 8, 1851, Dana to Bache, September 6, 1851, Bache to Henry, October 19, 1857, in Rhees Collection, Huntington Library; Gould to Frazer, November 22, 1856, Frazer Papers. See also Herman Fairchild, "A History of AAAS," *Science*, n.s. LIX (1924), 365–69, 385–90, 410–15.

Benjamin Gould was hired as director—on the payroll of the Coast Survey rather than the observatory itself.

By 1858, relations between Gould and the trustees were so bad that he was forcibly evicted from the premises. The public and later historians assumed that the fight was over pure versus applied science. The opposing forces, scientist versus citizen, seemed to confirm that suspicion. Further, Gould appeared to claim that Philistine trustees had blocked his every effort at research. However, the local citizens had generally a very sophisticated view of science, stemming partially from the donors' insistence that their bequest was not just for Albany or even the United States, but for the world. Gould's personality was much of the problem. In addition, there was some misunderstanding about the role of the scientific council in the management of observatory affairs.

The Lazzaroni showed a striking lack of real interest in the affair. Peirce, Bache, and Henry were reluctant defenders of Gould; their defense was in fact an argument of their rights as scientists to judge scientific work. None of the remaining five Lazzaroni rose to Gould's defense or rallied to the cause of pure science. Agassiz could have used his considerable influence in New York geological circles. Dana could have editorialized. Frazer could have worked through his friend Bishop Alonzo Potter, whose stature in Albany was almost as great as in Philadelphia, and whose debts to Frazer, Henry, and Bache were large. None used their powers, nor did anyone ask or expect that they would. In part, the reason for their inaction lay in the knowledge that Gould probably deserved what he got, but more important was the fact that the Lazzaroni were simply not that cohesive, that consistent, or that well-disciplined.[19]

---

19. See the previously cited collections of Lazzaroni papers, especially the Bache Papers, Vol. X, Library of Congress. See also *Inauguration of the Dudley Observatory at Albany, August 28, 1856* (Albany, N.Y., 1856); *The Dudley Observatory and the Scientific Council: Statement of the Trustees* (Albany, N.Y., 1858); *Defense of Dr. Gould, by the Scientific Council* (Albany, N.Y., 1858); B. A. Gould, *Specimens of the Garbling Letters by the Majority of Trustees of the Dudley Observatory* (Albany, N.Y., 1858); George N. Thatcher, *A Key to the "Trustees' Statement"* (Albany, N.Y., 1858); *An Address to the Citizens of Albany and Donors and Friends of the Dudley Observatory on the Recent Proceedings of the Trustees from the Committee of Citizens Appointed at a Public Meeting* (Albany, N.Y., 1858); and B. A. Gould, *Reply to the Statement of the Trustees of the Dudley Observatory* (Albany, N.Y., 1859).

POLITICS OF SCIENCE: NAS

It might be argued, of course, that the Lazzaroni were able to merge their scholarly tastes and strong personalities only in the face of what appeared to be a highly significant issue. That argument would assume that the Dudley Observatory fight did not seem important to the whole group. It also assumes that the whole group would unite for an important issue—for example, the question of a national academy for scientists.

No project was discussed among the Lazzaroni longer than the establishment of an academy or institute to shape public opinion with respect to science. Agassiz began calling attention to the need shortly after his arrival from Europe, and Bache pressed the issue in his AAAS presidential address of 1851. Throughout the decade, growing Lazzaroni dissatisfaction with the AAAS and with the local societies made the creation of some national elite corps of scientists more appealing. It is true that the four more ardent Lazzaroni—Peirce, Gould, Agassiz, and Bache—were the most potent scientific force behind the establishment of the National Academy of Sciences. It is important to note that these four carefully kept Henry ignorant of their plans. Further, three other Lazzaroni—Gibbs, Frazer, and Dana—were not parties to the creation. Neither their counsel nor their support was solicited by their clandestine comrades. Gibbs and Dana were each editors of the *American Journal of Science*. Surely loyal Lazzaroni would have used their editorial pens for favorable notice of the new academy. On the contrary, the journal pointed out that many worthy scientists were not among the original members and that their election in the future would be a greater honor than having been one of the cliquish incorporators.[20]

CONCLUSION

What conclusions can be drawn about that elusive group, the Scientific Lazzaroni? There were nine men who were conscious of

---

20. Agassiz to James Hall, January, 1851, in Clarke, *James Hall,* p. 200; Agassiz to Bache, December 22, 1854, Rhees Collection; "National Academy of Sciences," *American Journal of the Sciences and Arts,* 2d ser., XXXV (1863), 462–65.

having some special relationship to one another for social purposes only. Five or six members saw one another frequently and used those occasions to transact the business of science. Despite efforts by Gould and Peirce, the nine were never all together in one place for the explicit purpose of a Lazzaroni gathering. It is difficult to establish that more than six of them were ever assembled, even at the most gala meeting of the AAAS. One explanation is that the nine really had quite different interests. For example, there were twenty-six years separating the oldest Lazzarone from the youngest. Some functioned on a national level, while others, such as Frazer and Dana, rarely left home. Some were scientists of the first magnitude, while others were just beginning their careers, and one—Felton— was not a scientist at all, but a professor of classics. The point to be made in all this is that the Lazzaroni, as an organization, was not a primary affiliation for any of its members with the possible exception of its youngest, most fervent member, B. A. Gould. Each man had friendships outside the circle that were equally gratifying and useful from a business standpoint.

Why, then, were the Lazzaroni viewed as a cabal with coherent and disciplined points of view about the advancement of science? The allegation may have been rooted in those few occasions when some of the group did agree consistently on one issue. But such appearances are deceiving, and the contemporaries of the Lazzaroni were well deceived. More likely, large powers were ascribed to the Lazzaroni simply because some of its individual members were in positions of power. Men who are young and aspiring, or who have large egos and frustrated ambitions, in large measure create the conspiring establishments whose members they both fear and envy. Perhaps the Lazzaroni were created as much by those out of power as by those who were in. The existence of the group has been perpetuated by historians whose attractions to such a fascinating group magnified the significance of the scant evidence existing to support the allegations of conspiracy.

On the Flaws of American Physics: A Social and Institutional Analysis

In 1883 Henry Rowland of the Johns Hopkins University wondered "what must be done to create a science of physics in this country, rather than to call telegraphs, electric lights, and such conveniences by the name of science."[1] In the age of Edison, American physics was a low-grade activity. Though the profession doubled in size in the post-Civil War decades, by the late 1880s it consisted of no more than 150 men.[2] Though there were such distinguished American physicists as Albert A. Michelson, Willard Gibbs, and Rowland himself, physics in the United States neither ranked with its counterparts in Europe nor commanded much prestige among the general public. Inventors were the "scientific" heroes of the period.

Noting the popular supremacy of the inventor, historians have traditionally attributed the weakness of American physics in the late nineteenth century to a general national indifference to basic research. Calling that supremacy into question in his stimulating essay in this volume, Nathan Reingold, who considers the United States underrated in physics at that time, has contested the reality of the indifference itself. Both historical views rest on a doubtful premise—

---

1. Henry Rowland, "A Plea for Pure Science," in *The Physical Papers of Henry A. Rowland*, comp. A Committee of the Faculty of the Johns Hopkins University (Baltimore, Md., 1902), p. 594. In completing this chapter, I have benefited from the comments of the participants in the symposium on nineteenth-century American science held at Northwestern University.

2. Daniel J. Kevles, "The Study of Physics in America, 1865–1916" (Ph.D. diss., Princeton University, 1964), pp. 21, 121.

the assumption that one can speak of a general "American attitude" toward scientific research. After all, it is a commonplace among scholars that attitudes are hardly uniform across a large population. Just as political and cultural beliefs depend on social and economic status, as well as ethnic, racial, and religious identity, so attitudes toward science vary from group to group.

In the late nineteenth century, when less than 4 per cent of the nation had graduated from high school, a considerable fraction of the population was in fact indifferent to basic research. To cite the census figures of 1890, some 12 per cent of the country was black, and another 33 per cent foreign-born or the children of immigrants. Usually impoverished and uneducated, these Americans, amounting to almost half the population, were naturally unconcerned with scientific research, as were the native whites who were struggling against agricultural depression on the farms and high unemployment rates in the cities.

At the same time, of course, the nation had citizens who were not black, immigrant, or poor. Educated and well-to-do urban dwellers had flocked to hear Charles Lyell and Louis Agassiz lecture before the Civil War. If any group was likely to be concerned with the fortunes of science in the United States, it was this small group, whose money and interest in colleges exercised a large influence over the course of high culture in the country, including the culture of science. The question of whether the country was or was not generally indifferent to basic research is beside the point. Much more pertinent are the questions: Did the pre-Civil War patrons of science continue their interest after 1865? And, if they did, why was American physics not a good deal better?

THE CULTURAL PROMOTION OF SCIENCE

In the post-Civil War United States, science was in fact in vogue among Americans who considered themselves socially the best people. Members of the nation's mercantile and professional groups and of its town-and-country gentry made popularized science fashionable to an unprecedentedly intense degree. When the British physicist John Tyndall toured the northeastern cities in the winter of 1873 to speak on the marvels of light, they jammed the lecture halls to hear him and toasted him farewell at a glittering Delmonico's banquet. They read Edward Livingston Youmans' new *Popular Science Monthly,* which quickly won an audience as large as the *Nation's,* and pored over *Harper's* and the *Atlantic,* which both devoted numerous pages to the results of recent scientific research. At Dartmouth College in 1873 Whitelaw Reid, the editor of the New York *Tribune,* declaimed: "Ten or fifteen years ago, the staple subject here for reading and talk . . . was English poetry and fiction. Now it is English science. Herbert Spencer, John Stuart Mill, Huxley, Darwin, and Tyndall have usurped the places of Tennyson and Browning, Matthew Arnold and Dickens." [3]

The content of contemporary physics and natural history was provocative enough to inflame the interest of these Americans. Popularizers celebrated the theory of the conservation of energy— Youmans called it "the most important discovery of the present century"—and echoed Tyndall's sweeping elaboration on the mechanical theory of heat, that all the phenomena in the universe could be reduced to so many "modes of motion." [4] Numerous writers expounded the wonders of the theory of evolution, and some readers were infuriated. "If they believe that man descended from an ape," an irreconcilable minister snapped, "let them take a monkey . . . and, by a process of natural selection . . . , make a man out of him." The battle only amplified the attention Darwinism received. "I have never known anything like it," Youmans exulted. "Ten thousand *Descent of Man* have been printed and I guess they are all gone." [5]

Intellectual fireworks aside, the popularity of science drew considerable energy from the social drives of its patrons. They liked to consider themselves the nation's "most cultivated and intelligent

---

3. Quoted in Frank Luther Mott, *History of American Magazines,* 5 vols. (Cambridge, Mass., 1939–68), III, 105.

4. Edward L. Youmans, ed., *The Conservation and Correlation of Forces* (New York, 1869), p. v; C. W. Siemens, "Science in Relation to the Arts," *Popular Science Monthly,* XXII (November, 1882), 55.

5. John Trowbridge, "Science from the Pulpit," *Popular Science Monthly,* VI (April, 1875), 736; Richard Hofstadter, *Social Darwinism in American Thought* (Boston, 1955), p. 27.

people," as Youmans proudly described Tyndall's Manhattan audience, and by applauding science they set themselves apart in a country so exuberant about mere gadgets and machinery.[6] Raised in comfortable homes, they disdained postwar America's encomiums to material development and self-made men. "I took to pieces the claims of their practical men," Tyndall recalled his remarks in one of his American lectures. He was as "plain as . . . could be," yet his audience frequently interrupted with "loud approval."[7] The more the nation celebrated the advance of technology, the more its cultivated citizens insisted that its inventors were merely drawing on the treasury of science. The more the country measured progress in tons of steel and miles of track, the more they deplored the country's failure to contribute substantially to the scientific thought of the world.

As cultural nationalists, the patrons of science wanted to reform the colleges in order to make up for the nation's cultural deficiencies. Scientific studies sometimes demanded up to one-third of the undergraduate's time before the war, and professors of science accounted for a sizable fraction of the typical faculty. But the courses in physics, chemistry, geology, or mathematics were conducted more as instruments of liberal education than as professional training. Taught mainly by lecture and demonstration, they generally omitted laboratory work and were part of a prescribed curriculum which left no room for specialization.[8] Equally important, professors were not expected to advance knowledge, and graduate schools did not exist. When Charles William Eliot assumed the presidency of Harvard in 1869, he summarized "the failure of the system of the last forty years to breed scholars. . . . Asa Gray, Benjamin Peirce, . . . and Louis Agassiz are all going off the stage and their places cannot be filled with Harvard men, or any other American that I am acquainted with. This generation cannot match them. These men have failed to train their successors."[9]

---

6. *Popular Science Monthly*, II (February, 1873), 499.

7. Quoted in A. S. Eve and C. H. Creasy, *Life and Work of John Tyndall* (London, 1945), p. 171.

8. Stanley M. Guralnick, "Science and the American College, 1828–1860" (Ph.D. diss., University of Pennsylvania, 1969), *passim*.

9. Quoted in Henry James, *Charles William Eliot, President of Harvard, 1869–1909*, 2 vols. (London, 1930), II, 12–13.

In the post-Civil War decades, Eliot and his presidential peers elsewhere made the American system of higher education a good deal more of a breeding ground for scientists. Casting aside the prescribed curriculum, they encouraged specialization by gradually introducing the elective system and creating degrees in science. They also insisted that lectures and recitations be supplemented by laboratory work. Calling upon faculty to advance as well as to diffuse knowledge, they created graduate programs leading to the doctorate. By the 1880s a growing number of institutions were awarding Ph.D.'s in the sciences, and their leaders were consistently proclaiming the virtues of pure research, especially in science.

THE DIFFICULTIES OF ACADEMIC RESEARCH

Yet the fact remains that research—especially research in physics —did not flourish in the late-nineteenth-century university. Given the cultural nationalism of the college presidents and their public, it was clearly not because they held science in the same contempt as the vast majority of the country. Rather, it was because cultural nationalism was not the only reason that they had transformed the colleges in the first place.

To the patrons of science, educational reform was another weapon in the liberal assault on Grantism. Edwin L. Godkin's liberal-minded *Nation*—whose readers had a deep interest in the colleges —carried a good deal of educational material as well as denunciations of political practices. Writers in *Popular Science Monthly* matched their admiration for the triumphs of physics and natural history with indignation at high tariffs and paper money. What linked scientific training to the attack on corruption? Reformers actually expected, as an enthusiast announced, that "the chief influence of modern science upon . . . education will be its ethical influence." [10]

The college presidents of the day emphatically agreed. To them, America in the age of Ulysses S. Grant sorely needed high-minded men schooled in disinterested modes of thinking. Believing with

---

10. William P. Atkinson, "Liberal Education in the Nineteenth Century," *Popular Science Monthly*, IV (November, 1873), 23.

Charles William Eliot that the study of nature, especially in the laboratory, encouraged habits of "candid, fearless truth-seeking," and that science "ennobles and purifies the mind," they aimed to promote science as a powerful instrument of ethical training.[11] Like their predecessors, most considered the chief object of the college, even of the university, to be *the mind of the student,* not scientific discovery, not professional accomplishment."[12] Though Eliot and others proclaimed the virtues of research, in practice they downgraded research in favor of teaching.

Many professors of science were naturally at odds with presidential policy. With Rowland, some simply disliked having to teach "the ABC's of physics" to intellectual striplings.[13] More important, successful teaching did not normally win a scientist professional status. In any case, science was growing more complex. Had not the catholic "natural philosophy" turned into the diverse branches of "physics"? Scientists in discipline after discipline were forming their own professional societies and starting to publish their own journals. No scientist could hope to win a reputation in research without committing himself to the intricacies rather than to the generalities of his field.

If teaching failed to pay professional dividends, so did popularization. Almost no reputable scientist would consider airing a professional dispute before the public. (The law of gravitation, the Harvard astronomer Benjamin Gould once said, was hardly a matter for the democratic vote of a lay majority.) Professional reasons aside, many scientists simply disliked diluting their work for public consumption. Enough of them were contemptuous of popularization to make Youmans' *Monthly* bristle: "In order that the relations between science and the age may be what they ought to be, the world at large must be made to feel that science is, in the fullest sense, a ministry of good to all, not the private possession and luxury of a

---

11. Charles W. Eliot, *Educational Reform* (New York, 1898), p. 227; James, *Eliot,* I, 64.

12. Francis A. Walker to Alpheus Hyatt, August 29, 1889, Alpheus Hyatt–Alfred M. Mayer Papers, Princeton University Archives, Princeton, N.J.

13. Rowland to Daniel Coit Gilman, April 20, 1876, Henry A. Rowland Papers, Johns Hopkins University Archives, Baltimore, Md.

few, that it is the best expression of human intelligence and not the abracadabra of a school, that it is a guiding light and not a dazzling fog."[14]

Eliot and others were determined to keep science a guiding light for undergraduates and were able to do just that. So few jobs were available for physicists outside the academic world that most professionals in the discipline were professors. Moreover, as Lord Bryce noted, the American university president of the day held an "almost monarchical position."[15] He might confer with his faculty, but he exercised virtually autocratic power over salaries, appointments, promotions, and policy. With the exception of Daniel Coit Gilman at Johns Hopkins, the major presidents did little to ease pedagogical demands on a professor's time. They ran the university so that the advancement of knowledge paid little return in advancement through the college hierarchy; in fact, those who did no research often seemed to get ahead most swiftly. As Eliot admitted of Harvard, academic research in physics or in any other discipline required nothing less than "fanatical zeal."[16]

Combined with the widespread assumption of the economic irrelevance of pure science, the administrative antipathy to research placed academic physics under a financial handicap. It is true that the supporters of Johns Hopkins supplied Rowland with an ample budget and that philanthropists of a cultivated mind endowed new laboratories at Harvard and Yale. But as an observer noted with only slight exaggeration: "No university [has] the means of research . . . found in the shops of Thomas Edison at Menlo Park."[17] Academic physicists had to scrape along, wangle money where they could, often pay for apparatus from their own pockets. "Scientific

---

14. Gould, Presidential Address, *Proceedings of the American Association for the Advancement of Science*, XVIII (1869), 36; see also J. E. Hilgard to Alfred M. Mayer, March 21, 1877, Hyatt-Mayer Papers; *Popular Science Monthly*, XXXVIII (November, 1890), 121.

15. Quoted in George William Pierson, *Yale: College and University, 1871–1937*, 2 vols. (New Haven, Conn., 1952), I, 129.

16. Quoted in Samuel Eliot Morison, *Three Centuries of Harvard, 1636–1936* (Cambridge, Mass., 1936), p. 378.

17. *Electrical World*, VIII (November 20, 1886), 243.

men," Alexander Agassiz once wryly noted, "rarely spend their funds as their wives wish." [18]

However they spent their money, academic scientists could hardly fund the faculty and facilities necessary for good graduate work. One doctoral candidate went to Princeton and for a year, as he recalled, "browsed in the library, played in the laboratory, and deteriorated intellectually"; he quickly left for Europe.[19] Even the better-equipped universities attracted few graduate students. Most preferred to take their doctorates at the great scientific centers abroad. Between 1873 and 1890, American universities granted only twenty-two Ph.D.'s in physics.[20]

The academic situation not only hindered research; it also did a good deal to keep physics faculties small. With only a minute number of scholarships and fellowships available, most of the nation's youth could not afford graduate school, study in Europe, or even a college education. Even if one could pay his way through to a Ph.D., why go into physics when many professionals were saying that further progress could only be made by measuring constants to the sixth decimal place? In any case, as a scientist at Cornell remarked, "In this country, men devoted to science purely for the sake of science are and must be few in number. Few *can* devote their lives to work that promises no return except the satisfaction of adding to the sum of human knowledge. Very few have both the means and the inclination to do this." [21]

The physics community was in fact drawn from a narrow fragment of American society. Many in the discipline were the sons—or married the daughters—of well-to-do businessmen, farmers, lawyers, or ministers; almost all were white, Anglo-Saxon Protestants. If, as Rowland recognized, "the few excellent workers in our country must receive many accessions from without before they can constitute an American science," the number of all the workers increased very little because, in Rowland's day, physics was scarcely a career open

---

18. Agassiz to Miss Hyatt, August 3, 1900, Hyatt-Mayer Papers.
19. Quoted in William F. Meggers, "Henry Crew," *National Academy of Sciences Biographical Memoirs*, XXXVII (1964), 36.
20. See the table in Kevles, "The Study of Physics in America," p. 294.
21. W. A. Anthony, Section B Vice-Presidential Address, *Proceedings of the American Association for the Advancement of Science*, XXXVI (1887), 77.

to all.[22] The pursuit of physics was effectively restricted to the well-to-do.

THE PROBLEM OF COMPARATIVE QUALITY

Yet, despite the small size and academic difficulties of the profession, there were American physicists who did do research. And if, save for Willard Gibbs, none were creative theorists, many were deft masters of experimental technique. Unfortunately, few performed experiments of great importance to the science. Many American physicists seemed somehow inclined to follow fruitless trails and gather insignificant facts. The flaws of American physics ran deeper than the problem of limited opportunities in the discipline. American physics actually suffered a serious internal fault.

Professor Joseph Lovering of Harvard saw the fault in the profession's paucity of theorists. Lovering admonished: "Unless our physicists are contented to lag behind and gather up the crumbs which fall from the rich laboratories and studies of Europe, they must unite to delicate manipulation the power of mathematics."[23] Mathematics often did win victories where experiment failed. Theoretical analysis could suggest significant tests in the laboratory and help focus attention on important results which might otherwise be ignored. But Lovering's explanation was not satisfactory. Faraday hardly relied on abstruse mathematics; Rowland never used more than elementary calculus, and Michelson was uncomfortable with equations. Each was armed with acute insight into the problems of physics and each had a jugular instinct for what was likely to be a significant piece of work.

A more penetrating explanation of the initial weakness of American physics is George Daniels' argument that early-nineteenth-century American science, including physics, suffered from a naïve notion of Baconian empiricism. In Daniels' persuasive view, natural philosophers in the United States were overly attached to Bacon's

---

22. For the social composition of the profession, see Kevles, "The Study of Physics in America," pp. 114–16; Rowland, "A Plea for Pure Science," p. 610.

23. Lovering, "Mathematical Investigations in Physics," *Popular Science Monthly*, VI (January, 1875), 321.

assertion in *The Great Instauration* that the "true" approach to science pivoted on "keeping the eye steadily fixed upon the facts of nature." [24] Handed down from teacher to student, the Baconian tradition persisted past the Civil War and had a good deal to do with the continuing experimental emphasis of American physics. But even Daniels' argument fails to consider two points. Although many physicists seemed to confuse the importance of facts to science with a commitment to mere fact-gathering, their British colleagues were no less the heirs of Bacon; somehow they did good research. Equally important, the Baconian tradition was common to all the sciences in the United States, and not all of them were done as poorly as physics.

American geology was in fact distinguished, and one may profitably ask why such disparity in quality existed between the two disciplines. There were, of course, more geologists than physicists in the United States, but greater size is not necessarily a guarantee of greater quality. [25] On the intellectual side, members of both professions were engaged in a similar process of science, with the physicists testing the paradigms of thermodynamics and electromagnetism, and the geologists testing the theory of evolution. But if the numerical differences were not necessarily important and the intellectual states about the same, there was a key institutional difference between physics and geology in this period. While both disciplines were centered in the universities, geology also had the United States Geological Survey.

THE INSTITUTIONAL ADVANTAGE IN GEOLOGY

Created in 1879, the Survey quickly developed into a powerful agency under the leadership of its director, John Wesley Powell. No

---

24. George H. Daniels, *American Science in the Age of Jackson* (New York, 1968); Edwin A. Burtt, ed., *The English Philosophers from Bacon to Mill* (New York, 1939), p. 23.

25. In 1887 Section B, the physics division, of the American Association for the Advancement of Science had 145 fellows. In 1889 the new Geological Society of America had 187 members; in 1893, 224. See the list of fellows, *Proceedings of the American Association for the Advancement of Science*, XXXVII (1888), lxvi–lxxxvi; and the *Bulletins of the Geological Society*, I (1890), 586, and IV (1893), 450; also, Table 1 in Robert Bruce's essay in this volume.

laggard at Washington politics, Powell could usually find a job in the Survey for the relative of a well-placed congressman, and he was careful to distribute broadly the bureau's attractively, sometimes lavishly, illustrated publications around the capital. By 1884, the Survey's annual appropriation had jumped fivefold to $500,000, and increases continued through the rest of the decade. Powell spent the money to mount a multidisciplinary investigation of the evolution of the land. He employed topographers, geologists, and paleontologists and allowed the members of his diverse staff a good deal of latitude. Under Powell's imaginative administration, the U.S. Geological Survey was awarded the Cuvier Medal of the French Academy of Sciences.

The distinction of its work aside, the Survey was helped by Powell's administrative practices to become virtually the keystone in the institutional structure of American geology. Linked to Yale through staff members such as Othniel C. Marsh, the director of the Peabody Museum of Natural History, the Survey was also connected with numerous other academic institutions. Powell contracted for the services of sixty-nine consultants, including state geologists, mineralogists, chemists, and twenty-four professors of geology and allied sciences. The Survey's influence spread so widely through the country that Alexander Agassiz charged Powell with trying to dominate the discipline, even with wanting to make all the geologists in the United States his "satellites."[26] But Powell was not a dictator. He knew that, while scientists were eager to fit their individual labors into a "grand system . . . for the development of knowledge," they resisted direction from higher authority.[27]

Nevertheless, few, if any, geologists were as wealthy as Agassiz, a millionaire who could finance his own research and publications; most had quite mundane reasons to join Powell's purposeful national network. With the Survey's field expeditions and consultantships, they could better pursue their research. Access to the Govern-

---

26. Agassiz to T. H. Huxley, October 16, 1886, in *Letters and Recollections of Alexander Agassiz*, ed. G. R. Agassiz (Boston, 1913), p. 231.

27. U.S. Congress, Senate, Joint Commission to Consider the Present Organization of the Signal Service, Geological Survey, Coast and Geodetic Survey, and the Hydrographic Office of the Navy Department . . . , *Testimony*, Miscellaneous Document, 49th Cong., 1st sess., March 16, 1886, no. 82, serial 2345, p. 178.

ment Printing Office gave them a special publication organ to complement the *American Journal of Science*. The Survey had the loyalty of so many geologists in part because it provided professional opportunities not only for its own staff but for many other practitioners of the discipline in the country.

Consequently, the Survey was in a position to set intellectual standards of significance for American geologists. Such standards are the road signs of research; it is difficult to overemphasize what a crucial role they can play in keeping scientists on the highways—and out of the dead ends—of their discipline. Geniuses like Willard Gibbs can identify significant intellectual problems without institutional guidance. However, most scientists benefit from a system which establishes the questions to ask of nature, and even the techniques to use in answering them.

The leaders of the Geological Survey set standards by determining the task of its expeditions and establishing the editorial criteria of its publications. Because Powell was a good geologist and his associates were even better ones, the standards were high. Accordingly, whoever was paid or published by the Survey was discouraged from the gathering of insignificant facts. Equally important, the Survey, with its consultants spread through nineteen states and territories, made its influence felt in numerous academic localities. Its plentiful bulletins—over 100 had appeared by the mid-1890s—and its staff's consistent contributions to the *American Journal of Science* provided models of meritorious work for every American geologist. Under John Wesley Powell, in short, the U.S. Geological Survey not only funded a good deal of research but also shaped the intellectual thrust of the entire geological profession.

THE INSTITUTIONAL FLAWS IN PHYSICS

In the late nineteenth century, American physicists of course had nothing in Washington like the U.S. Geological Survey. There were some physicists in the Weather Service, others in the Coast Survey, and at least one in Powell's bureau itself, but none of these agencies dealt with the leading challenges of the discipline, such as electromagnetism, thermodynamics, spectroscopy, and statistical mechanics. In 1884, a committee of physicists and astrophysicists from the National Academy of Sciences urged the government to create a bureau of electrical standards and a "physical observatory" for spectral research.[28] The Academy's scientists claimed utilitarian value for their proposals, but physics had almost no friends in industrial circles to speak for it. As nineteenth-century liberals, the cultural patrons of science agreed with the *Nation:* "The Government should not keep a school for original research, and it is not advisable to see established at Washington laboratories of chemistry, physics, biology, etc., etc., intended only for such work as can be done elsewhere."[29] The government ignored the committee's proposal completely, and American physics remained without the kind of well-funded agency which could have guided the profession on a more intellectually significant course.

Academic physicists were left to private institutional devices. They had neither a national society nor a special journal of their own. They did, of course, participate in the country's multidisciplinary organs and organizations. But despite their names, the American Philosophical Society and the American Academy of Arts and Sciences exercised no more than local influence. More important, in the seventies and eighties even the most prestigious and genuinely national journals and societies were weak vehicles for setting standards for the profession of physics.

The *American Journal of Science* tried to maintain high standards for the articles it published in all fields, including physics. To judge the quality of the physics papers submitted, it seems that the chief editors usually relied upon an associate editor in the discipline; occasionally they called in outside consultants. When in 1873 the young and virtually unpublished Rowland submitted a bold experimental study of magnetic permeability, they turned to physicists at Yale, the home of the *Journal,* for an assessment of its merits. Following their advice, James Dwight Dana, one of the chief editors, rejected the paper, explaining to Rowland that he "needed more study before [he] could handle the subject satisfactorily." Rowland

---

28. For the report of the Academy's committee, see *ibid.,* p. 8.
29. "The National Government and Science," *The Nation,* XLI (December 24, 1885), 526.

actually needed no such further study. He sent the paper to the great British physicist James Clerk Maxwell, the world's leading authority on electricity and magnetism, and Maxwell was so impressed that he had it published forthwith in the *Philosophical Magazine*.[30] Evidently the physicists at Yale did not understand Rowland's study. Like even modern-day scientific periodicals, the *American Journal of Science* could miss an opportunity, as it did in this case, to advance the cause of high-quality physics simply because its expert consultants could be fallible.

But the problem of human fallibility is of course timeless; Dana's policy of maintaining high standards in physics foundered on difficulties that were inherent in the contemporary circumstances under which his periodical operated. While his associate editor or consultants might be considered distinguished physicists in the United States, most did not meet the criteria of high accomplishment prevailing in Western Europe. Most were probably not even well informed about the problems on the frontier of their discipline. (The physicists at Yale may not have recognized the importance of Rowland's study because, unlike Rowland, they had not read or understood Maxwell's *Treatise on Electricity and Magnetism*.) Generally mediocre themselves, the physicists upon whom Dana could call for judgments tended to set the *Journal's* standards for physics articles at a mediocre level. In any case, the *Journal's* proprietary editors, none of whom was a physicist, could not easily have insisted upon higher standards even had they been knowledgeable enough to do so. Unless the periodical published material in many disciplines, it risked the loss of subscriptions, which were virtually its sole source of financial support, and Dana was sure that a "purely geol[ogical] j[ournal] will not pay here." Compared to the number of papers in geology, few physics articles came along. To keep the contents of the *Journal* diverse, Dana promised the "early publication of any physical paper that may be received," which implied, however unintentionally, the publication of even mediocre studies.[31]

---

30. See Rowland's exultant note penned on the letter of rejection, James D. Dana to Rowland, July 2, 1873, Rowland Papers.

31. James D. Dana to Daniel Coit Gilman, June 20, 1884, Dana Family Papers, Stirling Library, Yale University, New Haven, Conn.

The American Association for the Advancement of Science had little ability to set standards for any discipline. By the 1880s, the AAAS had given physicists their own section, and at the annual meetings they exchanged ideas and read papers. But this subdivision was no more influential between meetings than the AAAS itself. The Association published only the proceedings of its annual meetings; its officers did little more throughout the year than plan for the next convention. In addition, it had only a minuscule research endowment. Physics had to compete with every other discipline for the available funds.

Some scientists considered the AAAS, fiscal difficulties aside, inherently unqualified to set standards of excellence. If the founders of the Association had been eager to marshall maximum support for science, they had been just as eager to create an authoritative organization which could explode the claims of scientific charlatans. ("Our newspapers are filled with puffs of quackery," Joseph Henry had complained, "and every man who can . . . exhibit a few experiments to a class of young ladies is called a man of science.")[32] To give the AAAS the widest possible base, its founders opened the membership not only to professionals but to interested laymen. To give the organization an authoritative voice, they wrote into its constitution such nominating procedures as virtually guaranteed the election of professionals to the Association's offices. Moreover, in 1874 a revamping of the document created a special class of members who were to be accomplished in research; under the new constitution, only fellows, as they were called, could hold office.

The AAAS may, in fact, have been oligarchic; however, a number of critics evidently considered it too democratic. Some ridiculed the Association for its nonprofessional membership.[33] Others were dissatisfied for a more fundamental reason. While the AAAS operated in a way that distinguished between professionals and laymen, its

---

32. Quoted in Howard S. Miller, "A Bounty for Research: The Philanthropic Support of Scientific Investigation in America, 1838–1902" (Ph.D. diss., University of Wisconsin, 1964), p. 11.

33. Benjamin Gould implied that some scientists cast "ridicule upon the Association" because it invited to its ranks "all lovers of science." Presidential Address, *Proceedings of the American Association for the Advancement of Science*, XVIII (1869), 36.

membership rules made no distinction between good and mediocre professionals. Standards of excellence could only be set by excellent scientists, and, in the opinion of the critics, an association which allowed any scientist to join was not in any position to encourage the best research.[34]

The critics' model scientific tribunals were the exclusive academies of Europe. Admission to the Royal Society of London or the French Academy of Sciences was possible only through election by their distinguished memberships. While the Royal Society numbered its members in the hundreds, only fifteen new members could be added to its rolls each year. The French Academy, with some fifty members, was even more exclusive. The two organizations appeared to affect significantly the life of science in France and England by publishing distinguished works and memoirs and by holding frequent meetings where first-rate scientists both informed and inspired their colleagues. Neither may have had large sums of money to dispense for research, yet both were seemingly quite influential outside the circles of their memberships, not least because they were elitist enterprises in societies characteristically deferential to official elites.

In America, advocates of an exclusive scientific tribunal looked to the National Academy of Sciences, a private body created by an act of Congress in 1863.[35] The charter of the NAS stipulated that it was to provide scientific advice to the federal government. But whatever the interest of its architects in achieving an organization for that purpose, they were also thoroughly aware of what an exclusive ward of the state might do for the country's science. Some of the founders hoped for a federal subvention. All undoubtedly expected that the Academy's official connection with the national government would endow it with weighty prestige. In any case, the

---

34. See, for example, Wolcott Gibbs to Othniel C. Marsh, May 15, 1882, Othniel C. Marsh Papers, Peabody Museum, Yale University, New Haven, Conn.; Simon Newcomb, "Exact Sciences in America," *North American Review*, CXIX (October, 1874), 300.

35. Alexander Agassiz to Simon Newcomb, May 30, 1906, Simon Newcomb Papers, Library of Congress, Washington, D.C.; Benjamin Gould to Newcomb, March 13, 1867, *ibid.*

charter of incorporation limited the membership to fifty. Louis Agassiz, one of the Academy's originators, rejoiced: "[Now] we have a standard for scientific excellence."[36]

Joseph Henry had not supported the move to create the Academy. He suspected that Congress would consider the idea "something at variance with our democratic institutions." Even if the Academy did win a federal charter, it would find getting appropriations impossible, and it might become "perverted . . . to the support of partizan politics." Moreover, Henry understood Americans well enough to know that an exclusive academy was bound to inject "continued jealousy and bad feeling" into the research community.[37]

When Henry became president of the Academy in 1867, he sought to remold it to the American context. He did not press for federal funds; he kept the Academy scrupulously out of politics and was instrumental in removing the ceiling on membership. Most important, perhaps, Henry paid shrewd attention to justifying the exclusive Academy in a democratic state. Patronage of research by rewarding success with titles and pensions was, contrary to the practice in Europe, unacceptable in America. But it was consistent with democracy to bestow honors for achievement, Henry argued. The creation of the Academy had opened in America another "avenue for the aspirations of a laudable ambition."[38]

Yet, despite Henry's modifications, in the seventies and eighties the NAS was largely ineffective in the cause of scientific excellence. The Academy could not overcome geography. British and French scientists were clustered within a few hours' travel of London or Paris and could easily attend the frequent sessions of the Royal Society or the French Academy. American scientists were dispersed over the entire eastern and midwestern United States, and Californians were hopelessly isolated. The National Academy met twice yearly at the most. Its sessions were poorly attended and had no

---

36. Agassiz to Alexander Dallas Bache, May 23, 1863 in *Science in Nineteenth Century America: A Documentary History*, ed. Nathan Reingold (New York, 1964), p. 210.
37. Henry to Louis Agassiz, August 13, 1864, *ibid.*, pp. 213–14.
38. U.S. Congress, Senate, *Report of the National Academy of Sciences, 1867*, Miscellaneous Document, 40th Cong., 2d sess., no. 106, serial 1319, p. 3.

more than a desultory effect on the tempo of American science.[39]

Much more important than geography were the Academy's inherent weaknesses. Henry had made it into an honor society. In the United States, could a principally honorific body expect to influence the course of research? Leo Lesquereux, a perceptive paleobotanist imported to America from Europe, argued that it could not: "What will you do with a body of scientific men who, in *America!* are expected to work for nothing and what influence can . . . the so called highest scientific society of the Land [exert], which cannot even publish a scientific memoir. In Europe, honor conferred is worth more than money but in America the same honor is worth nothing by itself." Henry's shunning of the public purse became a hallowed tradition. In the late nineteenth century, the Academy had only the income from a few private trust funds to dole out for the support of research. It could afford to publish little but, as a congressman once sneered, the obituaries of its deceased members.[40] Without money, the Academy could hardly give honor the leverage it needed to shape the development of American science.

The Academy's official status was of no help to it. While Othniel C. Marsh did mobilize the NAS to help Powell create the Geological Survey, on the whole, the leadership abided by Henry's admonition to maintain apolitical dignity. As the government acquired more of its own scientific bureaus, the advisory Academy became increasingly superfluous. By the late eighties, federal requests were so infrequent that some members were urging the leadership to volunteer the Academy's services. Marsh, the president of the NAS, refused. In his judgment, the Academy would "lose both influence and dignity by offering its advice unasked."[41] But, by allowing its official connection to become ever more nominal, the Academy devalued its prestige. In the United States, dignity without even the aura of power was not very commanding. Few scientists con-

---

39. See James D. Dana to Alfred M. Mayer, March 14, 1874, Hyatt-Mayer Papers; Alexander Agassiz to Simon Newcomb, May 30, 1906, Newcomb Papers.

40. Lesquereux to J. P. Lesley, March 21, 1866, in Reingold, *Science in Nineteenth Century America*, p. 222; U.S. Congress, House, *Congressional Record*, 45th Cong., 3d sess., February 18, 1879, p. 1564.

41. Quoted in Charles Schuchert, "Othniel C. Marsh," *National Academy of Sciences Biographical Memoirs*, XX (1939), 30.

sidered membership in the NAS important, Simon Newcomb noted, since it rarely brought "increased consideration in any quarters outside of the Academy itself."[42]

Above all, the National Academy of Sciences was a weak instrument because it was an honorific society in America. Though the limitation on the total size of its membership had been removed, it in fact remained an elitist organization since its constitution stipulated that only five scientists could be added to the rolls each year. As Henry had predicted, many nonmembers did resent its exclusiveness. Because the United States was so vast and plural a society, most could—and did—simply ignore it. In addition, the NAS annoyed a good many of its own members. With the Academy seated in the capital, federal scientists tended to dominate its affairs, particularly the elections. Although they may have differed with Alexander Agassiz's unmitigated opposition to Powell, scientists in Cambridge and New Haven did worry that the Academy was developing into a mere Washington clique.[43] As nineteenth-century liberals, they were antipathetic to machinations in the capital. As Americans, they were wary in principle of the centralization of science.

Rowland understood what it all meant for his discipline: Europe's great academies excited every physicist to his highest effort and provided him with "models of all that is considered excellent"; but America's National Academy was too weak to do the same.[44] Rowland might have added that the absence of any national institution which could set standards of excellence for his discipline was perhaps the principal flaw of American physics in the late nineteenth century.

---

42. Newcomb to Thomas C. Mendenhall, October 22, 1892, Thomas C. Mendenhall Papers, Niels Bohr Library, American Institute of Physics, New York, N.Y.
43. Agassiz to Othniel C. Marsh, May 4, 1884, Marsh Papers; see also Wolcott Gibbs to Marsh, May 15, 1884, *ibid*.
44. Rowland, "A Plea for Pure Science," p. 610.

The Beginnings of American Social Research

The conventional image of the social sciences in nineteenth-century America is one of speculative theory and utopian reforms. This view appears to be rooted in the self-concept of contemporary social scientists and the intellectual interests of present-day historians. During the first half of the twentieth century, American social scientists seemed determined to dissociate themselves from their nineteenth-century predecessors. In the name of disciplinary distinctiveness, methodological refinement, and theoretical purity, they tended to disown their legacy as merely one of grandiose systems and reformist impulses. Although this attitude of discontinuity with the past was doubtless necessary to the growth of the various disciplines, it led in time to a present-mindedness which blurred, if not obliterated, the actual achievements of earlier times.

If the grand theoreticians of the nineteenth century were dethroned by twentieth-century social scientists, they were enshrined by historians. Histories of the disciplines have been written, for the most part, in terms of the ideas of the theoreticians, great and minor, but rarely have the empirical investigations of social researchers received as much attention.[1] However, even a brief survey of research

---

This research was assisted by grants from the Case Research Fund and the American Philosophical Society, whose aid is gratefully acknowledged. It forms part of a larger study of the growth of American social research to be published by the University of Chicago Press. Reprinted in revised form from the *Journal of the History of the Behavioral Sciences* with permission of the publishers.

1. Exceptions are Luther L. Bernard and Jesse Bernard, *Origins of American Sociology: The Social Science Movement in the United States* (New York, 1943), and Joseph Dorfman, *The Economic Mind in American Civilization,* 5 vols. (New

activities reveals a wide range of investigations and an array of techniques. Ironically, it appears that there is more continuity with the present in the history of the methodology of social sciences than in the theory.

The activity to be examined here will be called social research to emphasize the predisciplinary nature of the enterprise. Social research refers to the systematic statistical study of the social characteristics and problems of various groups and populations. Today these studies fall mainly under the headings of demography, sociology, economics, epidemiology, and actuarial analysis, but in the antebellum nineteenth century they often shared the label of statistics. This overview will focus solely on the statisticians who concerned themselves with social matters and will exclude those who worked primarily with physical, biological, and economic phenomena. Also it will be limited to the pre-Civil War national period, which has been less thoroughly studied than colonial times or the late nineteenth century, for in this era there took place the first significant attempts at large-scale statistical work and the tentative beginnings of professionalization.

Between the Constitution and the Civil War, three broad streams of activity converged to shape the essential features of social research. The main stream of the statistical movement flowed from the administrative needs of the state and the economy and found expression mainly in the census and other official statistics. A second tributary, not entirely separate from the first, was the demographic stream of vital statistics. Stemming from the analysis of the facts of life and death, this current was given impetus by the epidemiological research of the medical profession and the demographic studies of insurance actuaries, and went beyond the numerical description of the regularities of birth, marriage, death, and disease in the search for broad natural laws. The third source of the statistical stream lay in the concern with social problems, and, following the usage of Adolphe Quetelet and the French statisticians, it became known as moral statistics. Originally focused on such problems as crime and

---

York, 1946–59); see also Paul F. Lazarsfeld, "Notes on the History of Quantification in Sociology," *Isis,* LII (1961), 277–333.

pauperism, it broadened to touch many social concerns and eventually appeared as "moral and social statistics" in official reports.[2]

In practice there was little clear-cut differentiation in the economic, demographic, and sociological emphases. The motives behind social research were as varied as the topics studied. The early social statisticians were moved by many concerns—national pride, scientific curiosity, economic calculation, administrative efficiency, reformist humanitarianism—which were often mixed in unexpected proportions. But running through the work is a pragmatic theme: statistics is a form of useful knowledge. And in the young Republic this was accolade enough.

FROM ENUMERATION TO SOCIAL SURVEY

The involvement of government in social investigation is most apparent in the development of the census. The Constitution provided only for an enumeration of "free persons" and "all other persons," but excluded "Indians not taxed" (Article I, Section 2). James Madison, however, urged a more elaborate undertaking. He argued that "they now had an opportunity of obtaining the most useful information for those who should hereafter be called upon to legislate for their country" and urged that the census bill be "extended so as to embrace some other objects besides the bare enumeration of the inhabitants."[3]

Echoing *The Federalist,* Madison proposed to gather basic economic data so that "the description of the several classes into which the community is divided should be accurately known." This would permit the Congress "to make proper provision for the agricultural, commercial, and manufacturing interests" in its legislation. He pointed out that "this kind of information has never been obtained in any country," and noted that "if the plan was pursued in taking

---

2. Harald L. Westergaard, *Contributions to the History of Statistics* (London, 1932); Helen M. Walker, *Studies in the History of Statistical Method* (Baltimore, Md., 1929); John Koren, ed., *The History of Statistics: Their Development and Progress in Many Countries* (New York, 1918); August Meitzen, *History, Theory, and Technique of Statistics,* trans. Roland P. Falkner (Philadelphia, Pa., 1891); and Vincenz John, *Geschichte der Statistik* (Stuttgart, 1884).

3. U.S. Congress, *Annals of Congress,* 1st Cong., 2d sess., 1790, p. 1077.

every future census, it would give them an opportunity of marking the progress of the society, and distinguishing the growth of every interest." Gathering detailed information "would furnish ground for many useful calculations," and would allow a check on the accuracy of the enumeration.[4] The House of Representatives agreed to have him draw up a schedule of questions. Madison expanded the constitutional provision by suggesting the classification of the population into five categories: free white males sixteen years of age or older; free white males under sixteen; free white females; free blacks; and slaves. The names of heads of families would also be obtained, and the census would classify the occupations of individuals under thirty occupational and industrial headings.

Although there were objections in the House to the questions on occupations, the bill was sent to the Senate as Madison had drafted it. But, as Madison wrote to Jefferson, "It was thrown out by the Senate as a waste of trouble and supplying materials for idle people to make a book."[5]

Although Madison had failed to convince his colleagues, the first census did move somewhat in the direction of his goal, for only the occupational questions were deleted. His argument that statistics were vital to legislation became the standard plea for the enlargement of the census, and his desire to monitor the growth of the new nation was repeated by scores of voices in the years to come. The topics which were added to the census schedules in the following fifty years are a rough index of the concerns of the young Republic and a crude measure of the effectiveness of the interest groups which besieged Congress.[6]

The intellectual community was the first to mobilize for the improvement of the 1800 census. The Connecticut Academy of Arts and Sciences, headed by the statistical enthusiast Timothy

4. *Ibid.*, p. 1078.
5. *Letters and Other Writings of James Madison,* 4 vols. (Philadelphia, Pa., 1865), I, 507; *History of Congress* (Philadelphia, Pa., 1843), pp. 191–93; and Edgar S. Maclay, ed., *Journal of William Maclay* (New York, 1890), pp. 194–95, 197–98.
6. Carroll D. Wright and William C. Hunt, *The History and Growth of the United States Census,* Senate Document, 56th Cong., 1st sess., 1900, no. 194, serial 3856, gives the details on the first eleven censuses but avoids mention of the political controversies involved.

Dwight, cooperated with the American Philosophical Society, whose president was Thomas Jefferson, in presenting memorials to Congress. Sharing Madison's views, Jefferson revealed in his memorial theoretical as well as practical aims for the census. In addition to restoring the deleted occupational questions, he wanted the age categories to be sufficiently detailed to construct life expectancy tables. He asked for data on nativity in order to estimate the contribution of immigration to population growth, a question of long-standing interest among intellectuals who concerned themselves with political economy. The census, he concluded, should provide a practical guide to public policy and, at the same time, might perhaps supply the basis for understanding economic laws.[7]

Congress ignored the memorials; its major innovation was putting the census under the authority of the Secretary of State. Thus the instructions for the second enumeration were drafted by Timothy Pickering, the field work was carried out under John Marshall, and the results were published by James Madison. However, the day-to-day supervision of the census appears to have been delegated by them to the principal clerk of the State Department, Jacob Wagner. The age categories were refined somewhat and age data were collected for free white females, but it was essentially a repetition of the first census. The meager results allowed Jefferson the Malthusian observation that American population was increasing in a geometric fashion, but little else could be determined.[8] Disappointment at the skimpy results was voiced by a number of writers including the energetic promoter of numerous schemes, Samuel Blodget, Jr., who in his *Economica: A Statistical Manual for the United States* (1806) gave suggestions for improvement. This curious work, rightly scorned

---

7. U.S. Congress, House, *House Report*, 41st Cong., 2d sess., no. 3, January 18, 1870, pp. 35–36; J. J. Spengler, "Malthusianism in Late Eighteenth Century America," *American Economic Review*, XXV (1935), 691–707; and E. P. Hutchinson, *The Population Debate: The Development of Conflicting Theories up to 1900* (Boston, 1967), pp. 258–319.

8. The census instructions, dated April 30, 1800, are in the Pickering Papers, Massachusetts Historical Society, Boston, Mass.; see also Irwin S. Rhodes, *The Papers of John Marshall: A Descriptive Calendar*, 2 vols. (Norman, Okla., 1969), I, 380, 384; and Andrew A. Lipscomb, ed., *The Writings of Thomas Jefferson*, 20 vols. (Washington, D.C., 1905), III, 330.

by Albert Gallatin as factually weak, was one of a number of attempts to chart the progress of the United States in statistical terms.[9]

By 1810, the demand for economic data had grown strong enough to move Congress, at the suggestion of Secretary of the Treasury Albert Gallatin, to authorize the collection of statistics on manufacturing in addition to population data. The census was directed by a political hack, John B. Colvin, a journalist-lawyer whose sycophancy was unbounded but whose statistical imagination was limited to relying on the instructions for the second census.[10] The economic data fared better in the hands of Tench Coxe, who produced an interesting report from sketchy returns. Dissatisfaction still reigned in statistical circles and led two congressmen, Timothy Pitkin and Adam Seybert, to independent efforts to order the statistical facts about the new nation. Seybert's *Statistical Annals* (1818) reflected careful original work and a discriminating evaluation of figures from official sources. Its superiority to available government publications is underscored by the fact that Congress subsidized its publication and widespread distribution.[11]

The 1820 census act combined the collection of population and economic data in the hands of Secretary of State John Quincy Adams, who chose to direct the work himself. In his methodical fashion he reviewed earlier census work and was appalled at the poor state of the census records and the lack of care in previous enumerations. Adams carefully drafted new instructions and expanded the schedule to include sex and age data on all classes of inhabitants.[12] For the first time, he instituted the collection of data on occupation and nativity. The latter inquiry revealed an interest

---

9. Everett S. Brown, ed., *William Plumer's Memorandum of Proceedings in the United States Senate, 1803–1807* (New York, 1923), p. 623; Wilhelmus B. Bryan, *A History of the National Capital*, 2 vols. (New York, 1914–16), Vol. I, *passim;* and *Dictionary of American Biography*, II, 380–81 (hereafter cited as DAB).

10. Charles Francis Adams, ed., *Memoirs of John Quincy Adams*, 12 vols. (Philadelphia, Pa., 1874–77), V, 130, 133–35; VI, 94–95, 288; and *National Intelligencer*, March 30, April 20, July 2, and July 9, 1810.

11. *DAB*, IV, 488–89 (Coxe); XIV, 639 (Pitkin); XVII, 2–3 (Seybert); U.S. Congress, *Annals of Congress*, 15th Cong., 1st sess., 1817–18, pp. 359–60, 2587; and *Annals of Congress*, 15th Cong., 2d sess., 1818–19, pp. 2549–50.

12. Adams, *Memoirs*, V, 125, 130, 133–35, 147–48.

in immigration and paralleled the efforts to collect similar data annually at the major ports of entry. At this time the War Department opened yet another field of social inquiry by beginning to collect vital statistics on the army.

The rising concern with social problems voiced by early reform movements found expression in the censuses of 1830 and 1840. Enumeration of the deaf and blind was included in the 1830 census, along with the refinement of age and occupational categories. By 1840, the census counted the illiterate, feeble-minded, and insane, and compiled as well a roster of military pensioners.[13] Thus in fifty years the census had grown into the instrument envisaged by Madison and Jefferson and had even moved beyond their original scheme.

Many groups with widely varied aims promoted the refinement of the census. Popular interest in statistics had been stimulated since colonial times by the newspapers, then the chief purveyors of such information, and this role increased greatly in the early nineteenth century. Joining them were the ever-popular almanacs, the best of which, *The American Almanac,* printed many valuable tables. But most important were journals such as *Niles' Weekly Register,* founded in 1811 by Hezekiah Niles, which from the beginning introduced statistics as a regular feature. Samuel Hazard launched his *Register of Pennsylvania* in 1826 with equal prominence given to statistics.

In 1839, two other publications heavily devoted to statistics appeared: Hazard's short-lived *United States Commercial and Statistical Register* and Freeman Hunt's more substantial *Merchants' Magazine and Commercial Review.* In 1846, a southern voice was added when J. D. B. De Bow launched his influential *De Bow's Review.* This period has been called "the era of almanac statistics" by one historian, but the stress on the American avidity for miscellaneous facts and spuriously accurate figures does an injustice to the quality of the statistical discussions which took place largely in the popular journals. Although the publications of scientific and historical societies are dotted with statistical articles, almost every major antebellum statistician is represented in the pages of *De Bow's Review*

---

13. Wright and Hunt, *History of U.S. Census,* pp. 89–96.

or the *Merchants' Magazine,* underscoring the strong ties of the early researchers to immediate problems and practical interests.[14]

Also concerned were the economists of the early national period. Although they seldom tested their theoretical assumptions with statistical research, they often touched on questions which demanded attention to social statistics. The question of population growth is a notable example, but the demand for adequate economic information led to a stress on systematic statistical work, which inevitably involved broader social inquiries.[15]

The business community, of course, was the chief source of pressure for improved commercial statistics. Since Hamilton's time, the Treasury had responded to demands for trade statistics, and gradually the department's mission was expanded to include an annual report on banking (1833) and a larger compilation on agriculture, commerce, and manufacturing (1845). By 1839 the Patent Office was collecting agricultural statistics. Not all these efforts fall under the narrower category of social research, but the demands of insurance companies for demographic data on which to base actuarial tables certainly may be included.[16]

Another vocal group had a professional stake in the same information. The concern with vital statistics among medical men in America had been strong since colonial times, but became intensified in the early days of the Republic when public health reform led to a series of epidemiological studies. In conducting sanitary surveys of community health and in pressing for the registration of vital statistics, medical men became major statistical collectors in addition to their more traditional activities, and they made up roughly half of the men who conducted creditable statistical studies before the Civil War.[17]

---

14. Bernard and Bernard, *Origins of American Sociology,* pp. 783–87.

15. Joseph J. Spengler, "On the Progress of Quantification in Economics," *Isis,* LII (1961), 258–76; and Dorfman, *Economic Mind, passim.*

16. Laurence F. Schmeckebier, *The Statistical Work of the National Government* (Baltimore, Md., 1925); and R. Carlyle Buley, *The American Life Convention: A Study in the History of Life Insurance,* 2 vols. (New York, 1953).

17. Wilson G. Smillie, *Public Health: Its Promise for the Future* (New York, 1955); George Rosen, *A History of Public Health* (New York, 1958); and Franklin H. Top, ed., *The History of American Epidemiology* (St. Louis, Mo., 1952). On the importance of medically-trained men in American science, see George H. Daniels, *American Science in the Age of Jackson* (New York, 1968).

The growing dissatisfaction with the federal census was perhaps greatest among those men who thought of themselves as professional statisticians. They could point out that by any standard the early censuses were amateurish operations, organized anew every ten years, enumerated by political partisans and compiled by squads of temporary clerks. There were abortive calls for a federal statistical bureau to centralize the uncoordinated efforts of the various agencies. As early as 1818, Hezekiah Niles noted that statistics had been "exceedingly neglected" in the United States, and suggested that a suitable person with a "liberal salary" be appointed in Washington to collect, arrange, and publish statistical tables.[18]

The two efforts to lobby for a permanent statistics office which came nearest success were led by Francis Lieber in 1836 and Zadock Pratt in 1844. Both became enmeshed in congressional politics and ultimately failed.[19] Another attempt to establish professional standards for the census was made by Archibald Russell of New York in his *Principles of Statistical Inquiry* (1839), the first book-length discussion of social research methodology published in the United States. Russell, a philanthropist born in Scotland and educated at Edinburgh, viewed the census in the Madisonian tradition and combined a perceptive exposition of techniques with a plea for extending the census into the realm of moral statistics.[20]

It was in Boston, however, that men interested in statistics first organized in an effective way. Late in 1839, the American Statistical Association was organized "to secure authentic information upon every department of human pursuit and social condition." The constitution and by-laws of the Association revealed the idealistic concerns which so often were mixed with practical considerations in the thought of the ante-bellum statisticians. "Another advantage which may be expected to flow from statistical inquiries in this country,

---

18. [Hezekiah Niles], "Statistical Works," *Niles' Weekly Register*, XIV (April 25, 1818), 142.

19. Frank Freidel, *Francis Lieber: Nineteenth Century Liberal* (Baton Rouge, La., 1947), p. 114; Pratt's plan is in U.S. Congress, House, "Bureau of Statistics and Commerce," *House Report*, 28th Cong., 1st sess., 1844, no. 301, serial 445.

20. See *Appleton's Cyclopedia of American Biography* (New York, 1891), V, 352; and *American Annual Cyclopedia . . . 1871* (New York, 1872), p. 573.

has relation to our peculiar civil institutions. It is of the utmost importance that, while the experiment of free political institutions is in progress, all of the facts pertaining to this experiment should be carefully gathered up and recorded. . . . In this way, if we succeed, it will be known *why* we succeed; if we fail, the *causes* of this failure will be apparent. . . . None of our institutions are in a perfect state. All are susceptible of improvements. But every rational reform must be founded on thorough knowledge." [21]

A similar line of thought prompted John Jay, one of the founders of the American Geographical and Statistical Society of New York in 1851, to write that for legislators, statistics are "a primer of practical knowledge" from which they can learn "the alphabet of legislative wisdom, and read easy lessons in political economy." Echoing the hopes of Adolphe Quetelet, he added that careful analysis of the census returns would be the basis for "educing from them natural laws, marked by mathematical accuracy, and possessing almost the certainty of moral truth." [22] Though this was more a pious hope than a program of research, the conviction that the laws of society would emerge from statistical tables was fervently espoused by a minority of American statisticians, notably Edward D. Mansfield of Ohio and Ezra Seaman of Michigan. [23]

These conceptions of statistics reveal the ideological matrix in which much of the discussion of the census took place. Statistics would provide the scientific basis for the art of government, the barometer of moral perfection, the ledger of economic progress, and the numerical record of the American experiment. Occasionally, the hope for an emergent science of society was voiced. But, in 1839,

---

21. *Constitution and By-Laws of the American Statistical Association* (Boston, 1840), pp. 3, 23; and Koren, *History of Statistics*, pp. 3–14.

22. John Jay, *A Statistical View of American Agriculture* (New York, 1859), pp. 6, 24. See also *The Charter and By-Laws of the American Geographical and Statistical Society* (New York, 1852); John Kirtland Wright, *Geography in the Making: The American Geographical Society, 1851–1951* (New York, 1952); and Paul J. Fitz Patrick, "Statistical Societies in the United States in the Nineteenth Century," *American Statistician,* XI (1957), 13–21.

23. See *National Cyclopedia of American Biography* (New York, 1909), XI, 206 (Mansfield); and Dorfman, *Economic Mind,* II, 952–55. Seaman's major work is his *Essay on the Progress of Nations . . .* (Detroit, Mich., and New York, 1846); Mansfield's thoughts are best seen in his annual reports as Commissioner of Statistics for Ohio from 1857 to 1867.

the American Statistical Association had a more immediate task: the promotion of better statistical work at local and federal levels. It had little influence on the census of 1840, and it was only after the shortcomings of that census were revealed that the Association began to play its role of a pressure group.

From the beginning, the census of 1840 was the focus of controversy. It had been supervised by a State Department clerk, William A. Weaver, a cashiered naval officer and sometime poet and pamphleteer, who had distinguished himself by compiling the *Diplomatic Correspondence of the United States of America*.[24] Congressional investigations into printing contracts were soon followed by serious questions as to the accuracy of the returns. Even before the census was printed, John Quincy Adams discovered inaccuracies in the enumeration of Negroes in Massachusetts. On close examination the printed tables showed clear discrepancies and internal contradictions, but, on the surface, they seemed to show that the rate of mental deficiency and insanity among free blacks in the North far exceeded the rate for slaves. Soon southern spokesmen, including John C. Calhoun, were citing the figures as proof of the humanitarian superiority of slavery. Protesting memorials were soon drafted by anti-slavery spokesmen, including Dr. James McCune Smith of New York. Smith, a leading Negro physician, educated at the University of Glasgow, was also moved to write an able series of statistical articles in rebuttal to southern claims. A memorial from the American Statistical Association, drafted by Dr. Edward Jarvis, effectively demonstrated the shortcomings of the census, but like the other petitions it failed to procure a correction of the enumeration.[25]

---

24. See *National Cyclopedia of American Biography*, XIII, 226 (Weaver); and Carl L. Lokke, "The Continental Congress Papers: Their History, 1789–1952," *National Archives Accessions*, No. 51 (June, 1954), pp. 1–19.

25. Albert Deutsch, "The First U.S. Census of the Insane (1840) and Its Use as Pro-Slavery Propaganda," *Bulletin of the History of Medicine*, XV (1944), 469–82; and Leon F. Litwack, *North of Slavery: The Negro in the Free States, 1790–1860* (Chicago, 1961), pp. 40–46. Also see Edward Jarvis, "Insanity Among the Colored Population of the Free States," *American Journal of Insanity*, VIII (1851–52), 268–82; and James McCune Smith, "The Influence of Climate on Longevity, with Special Reference to Insurance," *Merchants' Magazine and Commercial Review*, XIV (1846), 319–29, 403–18.

However, the debates over the defective returns focused needed attention on the disarray of official statistics. Secretary of the Treasury George M. Bibb expressed the view of a growing number when he charged that facts gathered by the census were "in a rude and indigested heap, inconvenient and inaccessible."[26] For the American Statistical Association, the skirmish led to a determination to see that the census of 1850 would be carried out in a thoroughly professional manner, but the cause of better statistics was not confined to the Association in Boston. In the South its most able advocates were James D. B. De Bow of New Orleans and George Tucker of the University of Virginia. *De Bow's Review* stimulated an interest in economic and social statistics along with its promotion of industrialization and the preservation of slavery. De Bow, an able practitioner as well as promoter of statistics, founded the first state bureau of statistics in 1848 and made an abortive attempt to introduce the subject in the college curriculum. Tucker was an equally forceful figure. During his twenty years on the faculty of the University of Virginia, he taught political economy with a heavy emphasis on statistics. His work, *Progress of the United States in Population and Wealth in Fifty Years* (1843), was widely read and adopted as a text in political economy.[27]

As the time for the seventh enumeration approached, the advocates of census reform began to make themselves heard. Early in 1848, memorials were sent to Congress by the New York Historical Society and the American Statistical Association. Drafted by Archibald Russell and Lemuel Shattuck respectively, the memorials decried the shortcomings of the previous census and advocated the use of statistical experts in planning the schedules for 1850. The weight of these opinions, together with the feeling in Congress that the 1840 census

---

26. *Report of the Secretary of the Treasury George M. Bibb,* Senate Document, 28th Cong., 2d sess., 1845, no. 21, serial 450, p. 3.

27. Ottis C. Skipper, *J. D. B. De Bow: Magazinist of the Old South* (Athens, Ga., 1958); Robert C. McLean, *George Tucker: Moral Philosopher and Man of Letters* (Chapel Hill, N.C., 1961); and Paul J. Fitz Patrick, "The Early Teaching of Statistics in American Colleges and Universities," *American Statistician,* IX (December, 1955), 12–18.

had been a national disgrace, led the Senate to pass a resolution supporting the idea of a small appropriation for planning, but the measure died in the House during the rush of last-minute business.[28]

The campaign was renewed in the next session, and its major spokesman in Congress was Senator John Davis of Massachusetts. A member of the American Statistical Association, he was familiar with the work of the Boston statisticians. He was in touch with Shattuck and had corresponded with Jesse Chickering and Nahum Capen, men whose views commanded attention. Shattuck had come into prominence as a founder of the statistical society, an effective advocate of vital statistics registration and the director of a census of Boston, and was soon to win even wider acclaim for his report on a plan for a sanitary survey of Massachusetts. Chickering, a Harvard-trained physician, had already made impressive studies of population growth and immigration. Capen, more a promoter than a practitioner of statistics, was a prominent publisher, who had once planned to launch a national statistical magazine in cooperation with the Association.[29]

Davis viewed their pleas for the utilization of statistical experts with sympathy, but he was also sensitive to the prerogatives of Congress. He therefore introduced a bill to create a Census Board—composed of the Secretary of State, the Attorney General, and the Postmaster General—which was empowered to appoint a secretary to solicit advice and draft the census schedules for submission to Congress for final approval. After much maneuvering and with the help of John Gorham Palfrey, the future historian, a House bill

---

28. *Proceedings of the New York Historical Society for the Year 1848*, pp. 45–47; see also Lemuel Shattuck to Charles Hudson, April 23, 1848, Hudson to Shattuck, May 18, 1848, and July 8, 1848, and William L. Dayton to Shattuck, August 11, 1848, Lemuel Shattuck Papers, Massachusetts Historical Society.

29. U.S. Congress, Senate, *Letters Addressed to the Hon. John Davis, Concerning the Census of 1849, by Nahum Capen and Jesse Chickering*, Miscellaneous Document, 30th Cong., 2d sess., 1849, no. 64, serial 533. See Lemuel Shattuck, *Memorials of the Descendents of William Shattuck* (Boston, 1855), pp. 302–12; *DAB*, XVII, 33–34; *National Cyclopedia of American Biography*, XII, 116; Joseph Palmer, *Necrology of Alumni of Harvard College 1851–52 to 1862–63* (Boston, 1864), pp. 50–51; *New York Journal of Medicine*, n.s., XVI (1856), 292; Francis S. Drake, ed., *Dictionary of American Biography* (Boston, 1872), p. 182; and Edmund Burke, "Nahum Capen," *Democratic Review*, XLI (1858), 397–412.

creating the Board was substituted for a bill which would have replicated the 1840 census with only slight modification.[30]

The Census Board did not select an experienced statistician as its secretary, but chose instead Joseph C. G. Kennedy, the former editor of two fervent Whig newspapers. Eventually it became apparent that his enthusiasm was no substitute for experience, so Archibald Russell and Lemuel Shattuck were called to Washington. As the first statistical consultants to the federal government, they performed their job with dispatch and quickly submitted a draft of the census schedules and part of the instructions. This draft, virtually unaltered, was rushed into print by Kennedy and the Board, much to the dismay of the consultants and the outrage of Congress. In a lengthy wrangle over the census bill which dragged on from January to May, Senator Davis defended the prerogatives of Congress by drafting his own schedules, and Russell and Shattuck exchanged letters voicing indignation at their peremptory treatment by Kennedy and his failure to give credit to them for their work. As Russell put it, "Certainly never did a public man get his work done by others more efficiently without a valuable consideration."[31]

When the census bill was finally passed the schedules were little changed from the original form submitted by Shattuck and Russell. Although the tensions generated in the controversy persisted, the utilization of statistical experts raised the federal census to a professional level and thus constituted a victory for the consulting statisticians.

Kennedy, too, won a victory, for he was appointed to superintend the census. Both he and James D. B. De Bow, who succeeded to the post after a change of administration, continued to draw on the advice of experts. Dr. Edward Jarvis, a Massachusetts physician who was prominent in statistical and medical circles for his work in vital

---

30. *Letters to John Davis*, p. 6; U.S. Congress, Senate, *Congressional Globe*, 30th Cong., 2d sess., 1849, pp. 626–29, 668–69.

31. See *DAB*, X, 335–36; *National Cyclopedia of American Biography*, XVI, 444; and J. C. G. Kennedy, *Progress of Statistics* (New York, 1861). See also U.S. Congress, Senate, *Report of the Census Board*, Executive Document, 31st Cong., 1st sess., 1850, no. 38, serial 558; Senate, *Congressional Globe*, 31st Cong., 1st sess., 1850, pp. 282–92; and Archibald Russell to Lemuel Shattuck, January 28, 1850, Shattuck Papers, Massachusetts Historical Society.

statistics, and Dr. Edward H. Barton, a New Orleans epidemiologist, worked on the mortality data from a medical perspective, while Levi W. Meech, a New England mathematician, constructed life expectancy tables from it.[32]

The census of 1850 quickly became the best-known product of American statistics. Kennedy's preliminary reports and De Bow's *Compendium* circulated in tens of thousands of copies, thanks to the bounty of Congress. De Bow's volume condensed the census findings and additional statistics into a relatively compact four hundred pages. Although De Bow tended to select figures which would cast slavery in its most favorable light, the data of the 1850 census provided Hinton R. Helper with most of the information for his *Impending Crisis of the South* (1857), the devastating indictment of the slave economy.[33] With the census of 1850, official federal statistical efforts reached the level of professional maturity and began to approach the goal Kennedy envisaged when he approvingly quoted the commissioners for the census of Ireland: "We feel, in fact, that a Census ought to be a social survey, not a bare enumeration."[34]

Social research in the early nineteenth century was not confined to the federal census, however. The other broad streams remain to be examined: the vital statistics movement and the efforts of social reformers.

THE FACTS OF LIFE AND DEATH

Many of the men who worked to improve the census were also active in the cause of vital statistics. They included physicians and

---

32. Gerald N. Grob, "Introduction," in *Insanity and Idiocy in Massachusetts: Report of the Commission on Lunacy, 1855,* by Edward Jarvis (Cambridge, Mass., 1971), pp. 1–71; William D. Postell, "Edward Hall Barton, Sanitarian," *Annals of Medical History,* 3d ser. IV (1942), 370–81; *Medical and Surgical Reporter,* III (1859–60), 36; Willard I. T. Brigham, *The Tyler Genealogy,* 2 vols. (Plainfield, N.J., 1912), I, 354–55; William G. Wallbridge, comp., *Descendents of Henry Wallbridge* (Litchfield, Conn., 1898), pp. 296–307; and *Historical Catalogue of Brown University, 1764–1914* (Providence, R.I., 1914), p. 157.

33. Hinton R. Helper, *The Impending Crisis of the South: How to Meet It* (New York, 1857), evoked a number of statistical rejoinders; see Hugh C. Bailey, *Hinton Rowan Helper, Abolitionist Racist* (University, Ala., 1965), chap. 5.

34. [J. C. G. Kennedy], *Report of the Superintendent of the Census for December 1, 1852, to Which is Appended the Report for December 1, 1851* (Washington, D.C., 1853), p. 148.

public health workers, whose contributions in establishing boards of health, lobbying for adequate vital statistics registration, and conducting sanitary surveys have often been noted. Less noted, however, have been the insurance actuaries whose interest in the statistics of life, death, and illness had a practicality of another kind.

The beginnings of all these interests lie in the colonial period, which has been thoroughly explored elsewhere. Colonial Americans had pondered the questions of longevity and the problems of epidemics. Additional impetus to inquiry was the running debate between Europeans and Americans regarding the salubrity of the American climate, the vigor of animal life, and the fecundity of the human population. To these questions were added the speculations of political economists, before and after Malthus, about population growth and the actuarial research stimulated by the rise of companies dealing with annuities and life insurance.[35]

During the Civil War, Edward Jarvis singled out several notable attempts to construct life expectancy tables based exclusively on American data.[36] The first of these was constructed by the Harvard theology professor Edward Wigglesworth. He had been interested for some time in the question of the rate of population growth in America and in 1775 had revised Franklin's early estimates. Faced with the more prosaic task of establishing an actuarial base for an annuity fund for clergymen and their widows, he constructed a crude life table from Harvard alumni records. Realizing its shortcomings, he enlisted the aid of the American Academy of Arts and Sciences in soliciting vital statistics in New England. From the records of 4,893 deaths, he eventually constructed a table which was published in 1793. The first table based on American mortality experience to be used by American insurance companies (although

---

35. James H. Cassedy, *Demography in Early America: Beginnings of the Statistical Mind, 1600–1800* (Cambridge, Mass., 1969).

36. See [Edward Jarvis], "On the Combination of Statistics for Determining the Average Rate of Mortality in the United States," *Statistics of the United States . . . in 1860* (Washington, D.C., 1866), pp. 517–24; Samuel Brown, "On the Mortality Amongst American Assured Lives," *Assurance Magazine*, VII (1858–60), 184–204; C. F. McCay, "American Tables of Mortality," *Journal of the Institute of Actuaries*, XVI (1870–72), 20–33; and Charles Gill, "Rates of Mortality in the United States and in California," *Assurance Magazine*, III (1853), 300–310.

most continued to rely on British tables), it was adopted by the Supreme Court of Massachusetts for the calculation of dower rights and life estates.[37] At the same time, Philadelphia lawyer William Barton, nephew and future biographer of David Rittenhouse, was engaged in similar work. Between 1789 and 1791, he presented three papers to the American Philosophical Society on estimates of population growth and the probabilities of the duration of life based on mortality data from Philadelphia.[38] Forty years later, Jonathan Ingersoll Bowditch, a son of Nathaniel Bowditch, published a life table for the white population based on the census of 1830. Bowditch, a mathematician, realized the limitations of tables constructed solely from mortality data or census enumeration, and, unlike Wigglesworth, he worked out a correction for population growth, which he estimated at about 3 per cent a year.[39] Another mathematician, Charles Francis McCay, a professor at the University of Georgia and consulting actuary, turned his talents to the question of high southern mortality rates, a problem which vexed northern actuaries and outraged southern policyholders. His table, published in 1850, was derived from a limited population; it was based on vital statistics of Baltimore.[40]

Discontent with the restricted scope of the previous studies led to an attempt to collect mortality data on a national scale in the census of 1850. Levi W. Meech constructed tables for male and female

---

37. Clifford K. Shipton, *New England Life in the Eighteenth Century: Representative Biographies from "Sibley's Harvard Graduates"* (Boston, 1963), pp. 507–17; Edward Wigglesworth, "A Table Showing the Probability of the Duration, the Decrement, and the Expectation of Life . . . ," *Memoirs of the American Academy of Arts and Sciences,* II (1793), pt. 1, 131–35.

38. David K. Cassel, *The Family Record of David Rittenhouse, Including His Sisters Esther, Anne, and Eleanor* (Morristown, Pa., 1896), pp. 16–24; Brooke Hindle, *David Rittenhouse* (Princeton, N.J., 1964), *passim; Proceedings of the American Philosophical Society,* I, 169, 181, and 193; and William Barton, *Observations on the Progress of Population, and the Probabilities of the Duration of Human Life, in the United States of America* (Philadelphia, Pa., 1791).

39. *National Cyclopedia of American Biography,* VI, 377–78; Gerald T. White, *A History of the Massachusetts Hospital Life Insurance Company* (Cambridge, Mass., 1955), pp. 63–71; J. Ingersoll Bowditch, "Tables Exhibiting the Number of White Persons in the United States at Every Age, Deduced from the Last Census," *Memoirs of the American Academy of Arts and Sciences,* n.s. I (1833), 345–47. Bowditch's table on dower rights is reprinted in *American Almanac, 1835,* pp. 83–91.

40. *DAB,* XI, 577–78; and C. F. McCay, "The Mortality of Baltimore," *Merchants' Magazine and Commercial Review,* XXII (1850), 35–44.

whites from the Massachusetts and Maryland returns, but under-reporting of deaths and population expansion due to the great immigration of the 1840s weakened his analysis. Despite the recognized shortcomings of the method, actuaries and physicians persisted in pressuring the census to collect mortality statistics and thus helped postpone adequate local registration of vital statistics until the twentieth century.[41] Life expectancy tables for Negro Americans were also constructed from the 1850 census. Returns from New England (free Negroes), Maryland (slaves), and Louisiana (mainly slaves) were analyzed for males and females and provided data for the continuing debate over slavery as well as an actuarial base for life insurance for slaves.[42]

The Massachusetts state census for 1855 was analyzed by Ezekiel Brown Elliott, a consulting actuary of Boston, who published his life table in 1857. This effort, too, was limited by the source of the data, but it is notable for the mathematical techniques Elliott employed to calculate his table.[43]

By the time of the Civil War, American life insurance companies had accumulated enough records to construct tables from policy-holder mortality experience. Charles Gill and Sheppard Homans, both mathematicians, utilized these data. In 1858, Homans, a Harvard-trained astronomer and mathematician, published a trial table which encouraged the American Life Underwriters conventions of 1859 and 1860 to advocate pooling records. From these sources Homans formed the American Experience Table in 1868, the first

---

41. [J. C. G. Kennedy], *Report of the Superintendent of the Census* . . . , pp. 10–13.

42. *Ibid.*, p. 13; on the problems of slave life insurance, see Eugene D. Genovese, "The Medical and Insurance Costs of Slaveholding in the Cotton Belt," *Journal of Negro History*, XLV (1960), 141–55; James H. Hudnut, *Semi-Centennial History of the New-York Life Insurance Company, 1845–1895* (New York, 1895), pp. 22–23; Mildred F. Stone, *Since 1845: A History of the Mutual Benefit Life Insurance Company* (New Brunswick, N.J., 1957), pp. 19, 41; Richard Hooker, *Aetna Life Insurance Company: Its First Hundred Years* (Hartford, Conn., 1956), pp. 12–15; and Terence O'Donnell, *History of Life Insurance in Its Formative Years* (Chicago, 1936), pp. 744–47.

43. E. B. Elliott, "On the Law of Human Mortality that Appears to Obtain in Massachusetts, with Tables of Practical Value Deduced Therefrom," *Proceedings of the American Association for the Advancement of Science*, 1857, pt. 1, 51–82; and Francis A. Walker, "Ezekiel Brown Elliott," *Proceedings of the American Academy of Arts and Sciences*, XXIV (1888–89), 447–52.

widely adopted table constructed on American data.[44] Edward Jarvis and Levi W. Meech also brought their experience to the analysis of the mortality figures from the 1860 census, but their efforts were curtailed by the Civil War.

The significance of the role of actuaries in the history of American social research is primarily methodological: they were the only researchers well enough trained in mathematics to bring relatively sophisticated techniques to bear on the analysis of the raw data of vital statistics. Physicians were largely responsible for stimulating the collection of the basic data. Bills of mortality had been published since colonial times, but the regular publication of detailed vital statistics was uncommon until such agencies as the Board of Health of Baltimore began issuing reports in 1815. Despite widespread interest among medical men in adequate registration, by 1833 only five major cities had effective registry systems, although information was recorded in a less satisfactory manner in many smaller towns.[45]

Laymen interested in public health allied themselves with physicians in the struggle for registration. Lemuel Shattuck and the American Statistical Association obtained a registration act in 1842 in Massachusetts, and in Rhode Island, Edwin Miller Snow led the fight for a similar bill, which was passed in 1852. Earlier, the army had begun to collect its own vital statistics under the leadership of Joseph Lovell, appointed Surgeon General in 1818.

Much of the concern with adequate vital statistics was due to the impact of epidemics and the climatological explanation of disease. This theory motivated General Lovell, and it particularly obsessed southern physicians. Among the men who produced statistical analyses exploring climate, morbidity, and mortality were James Wynne, Edward H. Barton, Daniel Drake, Bennet Dowler, and Charles McCay. James McCune Smith addressed himself to racial differences

---

44. Shepard B. Clough, *A Century of American Life Insurance* (New York, 1946), pp. 57–97; Homans' obituary, *Journal of the Institute of Actuaries*, XXXIV (1898–99), 144–47; and Gill's obituary, *Assurance Magazine*, VI (1855–57), 216–27.

45. John B. Blake, *Public Health in the Town of Boston, 1630–1822* (Cambridge, Mass., 1959); Robert Guttman, *Birth and Death Registration in Massachusetts, 1639–1900* (New York, 1959); and William T. Howard, Jr., *Public Health Administration and the Natural History of Disease in Baltimore, Maryland, 1797–1920* (Washington, D.C., 1924).

as well, while Nathaniel Niles and John Denison Russ studied comparative urban mortality. Isaac D. Williamson calculated sickness rates from the records of the Odd Fellows' lodge.[46]

The involvement of physicians in medical and social statistics was heightened by the sanitary surveys fostered by the American Medical Association (founded in 1847) and the four National Quarantine and Sanitary conventions (1857–60). Although scattered surveys had begun as early as 1806, the sanitary surveys of nine cities published in 1849 by the American Medical Association stimulated much interest in medical circles in health problems related to industrialization and urbanization. Lemuel Shattuck's *Report* of 1850 stemmed from this increased interest and was a landmark in the history of American public health.[47]

The physicians who participated in the sanitary surveys made a methodological contribution in exploring the techniques of collecting and analyzing data from large-scale surveys. Along with the growing cadre of men experienced in state and federal censuses,[48] the vital statisticians helped shape a research approach to the analysis of social problems in the ante-bellum period.

"EVERY RATIONAL REFORM"

The third stream of statistical activity which forms the social research tradition stemmed from the great burgeoning of reform

---

46. Percy M. Ashburn, *A History of the Medical Department of the United States Army* (Boston, 1929); *DAB*, XI, 440–41 (Lovell); XIII, 524 (Niles); XVI, 237–38 (Russ); XVII, 288–89 (Smith); William B. Atkinson, *The Physicians and Surgeons of the United States* (Philadelphia, Pa., 1878), pp. 361–62, 375; and Drake, *Dictionary*, p. 1010. See also Robert C. Davis, "An Early American Experiment in Health Insurance," *Cleveland Medical Library Bulletin*, Vol. XIV, no. 1 (1967), 4–12.

47. Harold M. Cavins, "The National Quarantine and Sanitary Conventions of 1857 to 1860 and the Beginnings of the American Public Health Association," *Bulletin of the History of Medicine*, XIII (1943), 404–7; the sanitary surveys are in *Transactions of the American Medical Association*, II (1849), 445–622; the best survey is by Josiah Curtis on the Boston-Lowell area.

48. Dr. Franklin B. Hough, who directed the census of New York state in 1855 and 1865 and of the District of Columbia in 1867, was perhaps the most able of these men; see *DAB*, IX, 250–52. See also, U.S. Bureau of the Census, *State Censuses: An Annotated Bibliography of Censuses of Population Taken After the Year 1790 by States and Territories of the United States*, comp. Henry J. Dubester (Washington, D.C., 1948).

movements that characterized the three decades before the Civil War. The "delinquent, defective, and dependent classes" of society claimed increasing attention, for the rapidly industrializing and urbanizing society soon overburdened the old systems of private charity and public assistance. However, not all reform activity resulted in empirical studies of social problems; utopian reformers seldom produced studies, but pragmatic reformers often engaged in research. Advocates of limited reform and incremental change frequently stimulated excellent studies which are strikingly contemporary in their focus and methods. As noted earlier, the federal government was gradually moved toward moral and social statistics. But it was on the local and state levels that most of the governmental action took place.

In New York, it was apparent as early as 1823 that a new state policy on poverty would require a solid empirical basis before legislation could be drafted. John Van Ness Yates, Secretary of State of New York, accordingly collected data on 22,111 recipients of public poor relief, which he analyzed to ascertain the apparent causes of poverty. He compared his findings with data from six other states and concluded that intemperance caused poverty for two-thirds of the permanent poor and for more than half of the temporary poor. On the basis of this survey the county poorhouse system, which was considered a notable reform at the time, was established.[49]

The role of alcohol in social problems is a recurring theme in the reform movements, but none turned correlation into causation more avidly than the advocates of temperance. Rarely did members of the dry crusade carry out research, but one exception was Samuel Chipman, who embarked on a survey of the inmates of the poorhouses and jails of New York state. His observations, first published in 1834, were extended to include data on inmates in four other states, and he drew the conclusion that intemperance was a major factor in the pauper and inmate populations he surveyed. Despite his propagandistic intent and simple methods, his findings left little doubt of the pervasiveness of alcoholism.[50]

49. David M. Schneider, *The History of Public Welfare in New York State, 1609–1866* (Chicago, 1938), 211–29.

50. Samuel Chipman, *The Temperance Lecturer* (Albany, N.Y., 1845); Bert L. Chipman, *The Chipman Family: A Genealogy of the Chipmans in America,*

Crime received much attention from reform groups and government alike. Before the Civil War, ten states (beginning with New York in 1829) collected criminal court statistics, and four states (beginning with Massachusetts in 1834) collected jail and prison statistics. These official figures on crimes and criminals are often incomplete and poorly tabulated, save for those in the reports of New York and Massachusetts. Supplementing official statistics were the reports of such groups as the Prison Discipline Society of Boston and the Prison Society of New York.[51]

The rivalry between two schools of penological thought stimulated the first real research in criminology. Attempts to compare the solitary-confinement system of Eastern State Penitentiary at Philadelphia with the congregate-and-silent system at Auburn State Prison in New York produced evidence which convinced some that the solitary system produced a higher rate of mortality, insanity, and recommitment. The questions of recidivism and rehabilitation were studied by a number of researchers; two other early studies stand out as fairly thorough. Gershom Powers attempted to follow up on all prisoners released from Auburn between 1825 and 1830 by writing to local district attorneys, postmasters, and law enforcement officials for reports on the adjustment of ex-convicts. He found that 112 of 160 men were decidedly steady and industrious and another 12 "somewhat reformed." The chaplain of Auburn, Benjamin C. Smith, later checked the records for the first 3,000 inmates admitted to that prison and found that 1 in 17 had been previously imprisoned at Auburn. Such studies are rare, and Dorothea L. Dix reported in 1845 that she knew of no prison which systematically followed up its former inmates.[52]

1631–1920 (Winston-Salem, N.C., 1920), pp. 110–11; and John Marsh, *Temperance Recollections* (New York, 1866), pp. 202–3.

51. Louis N. Robinson, *History and Organization of Criminal Statistics in the United States* (Boston, 1911); *Second Annual Report of the Prison Discipline Society, 1827* (Boston, 1827), *passim*, for examples of the statistical materials printed by such societies.

52. Negley K. Teeters and John D. Shearer, *The Prison at Philadelphia, Cherry Hill: The Separate System of Penal Discipline: 1829–1913* (New York, 1957), ch. 7; W. David Lewis, *From Newgate to Dannemora: The Rise of the Penitentiary in New York, 1796–1848* (Ithaca, N.Y., 1965); and O. F. Lewis, *The Development of American Prisons and Prison Customs, 1776–1845* ([Albany, N.Y., 1922]).

Another sector of the delinquent classes was studied by William Sanger, who surveyed New York prostitutes in 1855. Using policemen as interviewers, he collected data on 2,000 women. His survey is notable for its attention to the social backgrounds of the women. From the economic data he collected, he was able to make a rough calculation of the social cost of prostitution.[53]

Perhaps the greatest contribution of the reformers who concerned themselves with crime, intemperance, and poverty lay in their attention to the important concept that social ills produced indirect costs to the society at large in addition to the more obvious direct costs.

The "defective" classes, usually defined as the deaf, blind, feeble-minded, and insane, evoked compassion coupled with fact-finding. Samuel Gridley Howe, prominent for his work with the blind and deaf, produced an important study in his *Report on Idiocy in Massachusetts* (1848). However, the most extensive survey of retardation and insanity was carried out by Edward Jarvis. In 1854, as a member of the Commission on Lunacy, he sent out 1,690 letters, largely to physicians, and visited twelve institutions to ascertain the number of institutionalized and noninstitutionalized cases. Each case was rated as to the degree of manageability, the economic resources of the family, and the distance to the nearest appropriate institution. From his data he was able to assess the magnitude of the problem and the economics of its solution.[54]

Statistics on insanity and the contending claims of psychiatric schools raised a major ante-bellum dispute in medical circles. The reports of various asylums began to appear in the 1830s, and the *American Journal of Insanity,* beginning in 1844, published yearly summaries of this information. The ambiguous meaning of discharge and cure rates led such men as Edward Jarvis, Pliny Earle, and Isaac Ray to make critical analyses of the statistics. Eventually, Earle summarized his statistical labors in *The Curability of Insanity*

---

53. William Sanger, *The History of Prostitution: Its Extent, Causes, and Effects Throughout the World* (New York, 1858); see also *Appleton's Cyclopedia of American Biography,* VII, 240.

54. Harold Schwartz, *Samuel Gridley Howe, Social Reformer, 1801–1876* (Cambridge, Mass., 1956), chs. 5–7 and 9–10; and Edward Jarvis, "Insanity and Idiocy in Massachusetts," *American Journal of Insanity,* XII (1855–56), 146–82.

(1877), which demolished the optimism of the early mental health workers.[55]

Increasingly in the 1840s and 1850s the energies of many of the reformers were channeled into the controversy over the "peculiar institution," but the question of slavery generated more rhetoric than research. Calhoun's use of the 1840 census data on Negro insanity rates was symbolic of the difficulty of statistical discourse whenever slavery was the issue. As we have seen, demographic studies of regional and racial mortality rates also became involved in the controversy. There were few studies of the social and economic conditions of slaves or free blacks that explored intensively the topics which were debated so extensively; however, some studies of the conditions of free northern Negroes were creditable research performances. As early as 1831, the first Annual Convention of the People of Color, in Philadelphia, resolved to study the conditions of free Negroes, but the best surveys of the black communities were carried out in Philadelphia by the Pennsylvania Society for Promoting the Abolition of Slavery and the Society of Friends. Published between 1838 and 1856, the results of these surveys clearly revealed the patterns of residential segregation, educational deprivation, and occupational limitation characteristic of the northern cities. Scattered studies of this kind were carried out in Boston, New York, and Cincinnati as well, and less complete statistics on other towns are to be found in the Negro periodicals and abolitionist journals.[56]

Immigration was another topic of controversy, especially in the

---

55. Norman Dain, *Concepts of Insanity in the United States, 1789–1865* (New Brunswick, N.J., 1964). Representative articles are Isaac Ray, "The Statistics of Insane Hospitals," *American Journal of Insanity,* VI (1849–50), 23–52; Edward Jarvis, "On the Supposed Increase of Insanity," *American Journal of Insanity,* VIII (1851–52), 333–64; and Pliny Earle, "On the Curability of Insanity," *American Journal of Medical Sciences,* n.s. V (1843), 344–63.

56. See, for example: *The Present State and Condition of the Free People of Color of the City of Philadelphia* (Philadelphia, Pa., 1838); and *Statistics of the Colored People of Philadelphia* (Philadelphia, Pa., 1856). Cf. J. L. Dawson and H. W. De Saussure, *Census of the City of Charleston for the Year 1848 . . .* (Charleston, S.C., 1849); and Joseph Bancroft, *Census of the City of Savannah, Georgia . . . ,* 2d ed. (Savannah, Ga., 1848); and [James McCune Smith], "A Statistical View of the Colored Population of the United States—From 1790–1850," *Anglo-African Magazine,* I (1859), 33–36, 65–69, 97–100, and 140–44.

North. Adam Seybert made his own estimate as early as 1817, as did George Tucker and Henry C. Carey somewhat later. The most careful analysis was made by Jesse Chickering, who estimated the immigration from 1800 to 1848. Official statistics on immigration were published by the federal government without engendering the degree of acrimony which attended the slavery issue.[57]

This brief sampling of the research efforts of various reform groups can only indicate the range of investigations. In general, social research was carried out mainly by those reformers who aimed at specific institutions or who sought to influence government to create legislation on concrete social problems.

THE END OF THE BEGINNING

From the outset, stress on the utility of statistical knowledge gained support for statistical endeavors from government, business, and reform groups alike. The belief that the new American institutions were amenable to increasing perfection gave further impetus to a national self-examination that was more than self-congratulation. Although policy did not automatically emerge from tables of data, the first attempts to shape legislation and justify reform on the basis of statistical evidence did take place in the ante-bellum period. On the local level such endeavors as the public health movement rested much of their lobbying on a statistical basis, while national issues such as the tariff, slavery, economic growth, and immigration were debated increasingly in statistical terms. Occasionally, as in the regulation of the insurance industry, men like Elizur Wright employed statistical analysis directly in shaping public policy.[58]

Measured in European terms, the American performance may be seen as weakest in the failure to create a federal statistical bureau. But while the states' rights philosophy militated against such a cen-

---

57. Arnold W. Green, *Henry Charles Carey, Nineteenth-Century Sociologist* (Philadelphia, Pa., 1951); Jesse Chickering, *Immigration into the United States* (Boston, 1848). Carey's estimates in *The Slave Trade, Domestic and Foreign* (Philadelphia, Pa., 1853) are still useful; see Philip D. Curtin, *The Atlantic Slave Trade: A Census* (Madison, Wis., 1969), pp. 3–13, 72–75.

58. Massachusetts Insurance Commissioners, *Massachusetts Reports on Life Insurance: 1859–1865* (Boston, 1865) is mainly Wright's work.

tralized agency, it often encouraged excellent statistical work on the state and local level. In the absence of any enduring attempt to educate statisticians in colleges and universities, the practical experience gained in city and state surveys was the major source of training for nearly all the ante-bellum professional statisticians. Moreover, the establishment of statistical societies helped spread the knowledge of techniques while it offered an organized base for promoting improvements.

Contact with European statisticians increased throughout the period and helped stimulate activities on the American scene. Adolphe Quetelet's honorary membership in both American statistical societies symbolically linked European accomplishments with the newer endeavors in the United States. Furthermore, statistical influence flowed both ways. British advocates of a decennial census consulted the American experience. The many foreign travelers who described the American scene drew heavily upon the native statisticians and made them known abroad. Honors, too, were distributed internationally. In 1837, Pitkin was awarded a medal by the Société Française de Statistique Universelle for his volumes on economic statistics. Kennedy, touring Europe as Superintendent of the Census, spoke before various statistical groups and was one of the founders of the International Statistical Congress. Shattuck carried on a correspondence with most of the statistical leaders of Europe, and his work was widely lauded by them. Jarvis attended the 1860 meetings of the International Statistical Congress and was active in the first effort to establish a uniform classification of causes of death. A corresponding member of the British National Association for the Promotion of Social Science, he was later a founder of the American Social Science Association in 1865.

Statistics at this time usually implied numerical description rather than mathematical analysis, but it is clear that Americans who worked with life insurance problems were conversant with emerging actuarial methods. Although the United States produced no individual statistician comparable to Quetelet, in the actuarial sector, at least, it developed a small cadre of mathematically trained men whose talents equaled their European counterparts. However, most men who considered themselves professional statisticians confined their endeavors to elucidating practical problems through descriptive statistics.

These practical statisticians were concerned with knowledge of social conditions which had direct applicability in the formulation of public policy. Their philosophy of science stressed induction over deduction. They saw quantitative methods as the soundest approach to valid and reliable knowledge. When they ventured into the realm of theory, they shied away from grand systems and preferred generalizations of a more modest scope. Far from being scientific purists, they wanted their findings to be applied to immediate problems. When they associated themselves with reform movements they leaned toward pragmatic, incremental improvements rather than all-embracing programs for social change.

Thus, in contrast to the image of grand theory and utopian schemes which has tended to dominate our view of the nineteenth-century social sciences, we find a vigorous empirical and pragmatic strain which, in many ways, has more continuity with the present than does the heritage of nineteenth-century social theory.

III

Science, Technology, and Nineteenth-Century Entrepreneurship

Science, Technology, and Economic Growth: The Case of the Agricultural Experiment Station Scientist, 1875–1914

Clichés of historical explanation are not always wrong; they are simply not very useful. A familiar example urges the importance of the relationship between science, technology, and economic development in the historical evolution of Western society. Can one object? The problem, of course, lies not in creating a willingness to concede this general point, but, given the present state of the art, in understanding this relationship in particular contexts and at particular moments in time. Despite its obvious importance, students of American society have devoted little attention to the historical interaction between science, technology, and social context; aside from a handful of case studies, the historical literature is spotty and unsystematic.

This case study of the scientists employed by American agricultural experiment stations between the 1870s and 1914 is designed to explore the interaction. Its central theme is the behavior of scientists and scientist-administrators within an institutional context defined by social and economic factors. The first part of this chapter will be a brief outline of the nature of lay expectations—a primary dimension of the scientist's reality—followed by a characterization of the response of scientists to what was originally a far from satisfactory

This chapter appeared in a somewhat shortened version as "Science, Technology, and Economic Growth: The Case of the Agricultural Experiment Station Scientist, 1875–1914," *Agricultural History,* XLV (January, 1971), 1–20. The arguments presented here are from a larger study of the experiment station movement in the United States.

work situation; the final pages argue that the experiment station scientist's efforts to shape a professionally satisfactory environment served as a component of economic and institutional change.

THE SOCIAL CONTEXT

In 1875, the state of Connecticut established an agricultural experiment station. North Carolina and New York followed suit in 1877 and 1880 respectively, as did a number of other states and land-grant colleges in the early 1880s. In the Hatch Act of 1887, the federal government made the experiment station a national institution by providing $15,000 a year for the support of a station in each state.[1] The motivations of the state and national legislators who approved these measures seem transparent enough. Easily clothed in the neutrality of science and justified in terms of the traditional virtues granted the yeoman cultivators of the nation's farms, these subventions for research were pork-barrel concessions to farm power. The stations would help the farmer adjust to an increasingly competitive world market and would rationalize and systematize his operations—would provide, that is, a seemingly conservative alternative to socialist or Farmer's Alliance schemes for adjusting to changed economic and demographic realities.

The desire of politicians and farm spokesmen for scientific aid was real enough, but their conception of the scientist's role was dismayingly imprecise. Although since the 1850s professional scientists had been leaders in the campaign of "education" and agitation in favor of the experiment station idea, such scientists had found it both difficult and impolitic to be candid in their predictions of what an experiment station might be and do.

To many interested farmers and journalists, nevertheless, the role of the experiment station seemed all too clear: the station was to perform the experiments in cultivation which the individual farmer, lacking time and opportunity, could not; common sense, order, and

---

1. The best general history of agricultural research in the United States is still A. C. True, *A History of Agricultural Experimentation and Research in the United States, 1607–1925,* USDA Miscellaneous Publication No. 251 (Washington, D.C., 1937).

precision allied with the American's native ingenuity were qualities adequate to the task. This was stated explicitly in the writings of farm journalists and legislators who enthusiastically supported the experiment station movement.[2] A good proportion of the station's nonprofessional advocates, as well as its self-appointed critics, expected—even in the twentieth century—that the experiment station would be operated as a model farm, indeed a model farm which would show a profit on its operations. (How, after all, could it be presumed a model if it lost money?) Up through the First World War, experiment station workers had to answer the criticisms of visitors who could not understand why a stand of wheat might be dwarfed and why scores of horticultural varieties were not to be seen in vigorous profusion. Such conceptions had a long and tenacious history. J. B. Turner, for example, a putative father of the land-grant college, argued in his 1848 plan for an industrial university that the professor of agriculture should be judged by his ability to plan experiments in which no money would be lost. "I would put no public funds into the professor's hands (certainly none beyond the original outfit) to squander in any day dreaming or absurd speculations. I would have every new experiment bear directly on his own private purse."[3]

Only with respect to chemistry did the "intelligent farmer" assume that special training might be a prerequisite for the experiment station staff member. Indeed, to many Americans agricultural science *was* chemistry; Liebig and his doctrines were familiar even to remote farm communities. But even with respect to chemistry, the experiment station scientist found that assumptions underlying the expectations of his farm constituency implied career problems. One difficulty was the popular illusion that simple testing procedures could ensure soil fertility; once missing constituents were identified

---

2. "The average farmer," as the *Rural New-Yorker* editorialized in 1886, "cannot afford to experiment in a careful systematic way for himself. Hence the value of Experiment Stations, if only they be conducted by intelligent, earnest, practical men" (XLV [August 21, 1886], 548).

3. J. B. Turner to Jonathan Blanchard (1848), Turner Papers, Illinois Historical Survey, University of Illinois, Champaign, Ill. Cf. Winton U. Solberg, *The University of Illinois, 1867–1894: An Intellectual and Cultural History* (Urbana, Ill., 1968), p. 135.

in the test tube, the farmer need only apply the prescribed fertilizer, and a marginal farm would become a source of profit. Throughout the last decades of the nineteenth century and even into the first years of the twentieth, farmers confidently sent in samples of soil for analysis and requested individually tailored advice for purchasing fertilizers.

The marketing of chemical fertilizers provided another ambiguity for the first generation of station scientists. Perhaps the most telling single argument for the creation of experiment stations in the late 1870s and the 1880s lay in their potential for regulating chemical fertilizers. Station supporters in older farming areas, conscious of competition from newer and more fertile states and indignant at the sometimes casual ethics of fertilizer manufacturers and dealers, assumed that experiment stations would act as "fertilizer control" agencies. (The North Carolina and Maine stations actually bore this designation as part of their formal titles.) Experiment station advocates were willing enough to cater to these desires; this public was a source of leverage in gaining the support of influential farm and business leaders. Yet once the stations were established, the young chemists who staffed their laboratories often found themselves prisoners of a deadening routine of fertilizer analysis.[4] In general, indeed, research and development were never clearly distinguished in the public mind from regulation and inspection.

In addition to sharp consciousness of economic need and vague public expectations in regard to science, more general ideological factors played a role in shaping the experiment station scientist's environment. Perhaps the most important factor in setting intellectual guidelines was the generally assumed and morally absolute vision of an intelligent, prosperous yeomanry as a necessary basis for an enduring democracy. Clearly the personal aspirations of station scientists were secondary to the well-being of this virtuous class. Experiment station scientists in the years before the First World War

---

4. C. W. Dabney to Mrs. Dabney, March 16, 1881, Dabney Papers, Southern Historical Collection, University of North Carolina, Chapel Hill, N.C.; A. T. Neale to W. O. Atwater, March 13, 1882, and March 17, 1884, Atwater Papers, Edgar Fahs Smith Collection, University of Pennsylvania (originals at Wesleyan University, Middletown, Conn.); *First Annual Report of the Connecticut Agricultural Experiment Station, 1876,* p. 353.

thus had to deal with a constituency aggressive in its rectitude and casual in its assumption of the right to make and enforce demands upon the performance of station scientists. Chemists and botanists were seen as public servants. Government in the nineteenth century was not something of which to stand in awe, but something to despoil. Thus, for example, the plaintive complaint of experiment station directors that visitors thought it their right to pick anything growing on the "state farm." The fruit belonged to the people![5]

The station scientist not only faced arbitrary and inconsistent public demands, but he was, as well, outside the established structure of administrative authority. Scientists or scientist-administrators were newcomers at the mercy of a sometimes unsympathetic academic hierarchy of college presidents and faculties. (The Hatch Act, which endowed the stations in 1887, required that they be part of the land-grant college, except in those states where an independent station already existed.) Imposed from above by federal fiat and at first unprotected by federal regulation and control, the station and its budget were at the mercy of existing—and almost always underfunded—academic administrations. A responsive and appropriate context for scientific research and even the adoption and dissemination of empirical findings would require power within each state's hierarchy of economic and political power; this power could then be utilized in the academic sphere as well.

Some of the problems faced by experiment station scientists were not peculiar to the stations but were generally characteristic of the American academic world in the late nineteenth century. The pressure of an aggressive nonprofessional constituency seemed particularly intolerable in a period during which ambitious young scientists were increasingly influenced by the German research ideal. The intellectual and behavioral demands of the several scientific

---

5. This attitude was not limited to fruits and vegetables. When the *Sacramento* (Calif.) *Record-Union* printed Professor E. W. Hilgard's lectures without permission, the newspaper's manager argued that such lectures were not Hilgard's private property: "He is paid a salary by the university in behalf of the people of the state of California. . . . I believe a student has the right to take these lectures . . . and furnish the same for publication." W. H. Mills to Rev. J. H. Bonte, January 30, 1882, Hilgard Papers, Bancroft Library, University of California, Berkeley, Calif.

disciplines made it difficult for the scientist who defined his aspirations in disciplinary terms to accept the constraints of an institutional context dominated by the economic needs and social preconceptions of this constituency. Most station scientists had college-teaching assignments as well as station duties and, in this capacity, were expected to teach a multitude of courses for long hours at low wages. A combination of German, chemistry, and botany—with certain "pastoral" duties as well—was not atypical.[6] Particularly frustrating, especially to that increasing number of American scientists exposed to the values of the European academic disciplines, was the universal assumption that research was not a part of the college professor's expected duties. Standards other than competence in one's field were still routinely applied; even chemists, for example, found religious criteria important in securing a college position.[7]

Southern institutions demanded not only piety but Democratic orthodoxy—at salaries even lower than those offered in the North. As late as 1889, for example, Texas hoped to hire a "competent analytical chemist" for $800–900. An applicant for a position as professor of natural history assured the president of the University of Kentucky that: "Your last three requirements are perhaps the easiest met. I am a *Southerner,* intensely so, both by birth and education; a *democrat,* because I could be nothing else, since I was born of democratic parents, and in a democratic state, and nourished on democratic pabulum." The gentleman concluded wanly that he was a "passive" Methodist. While it was well known that Southern institutions hesitated to hire Northerners, Northerners felt great hesitancy in accepting Southern positions.[8]

---

6. S. M. Babcock, agricultural chemist at the Geneva Experiment Station, received, for example, an offer for such a position in 1884 from Lombard College in Galesburg, Illinois: "We want some one that can teach Chemistry, Physiology, Botany, Zoology, Geology, German, &c. The number of recitations will be about five per day." J. V. N. Standish to Babcock, July 22, 1884, Babcock Papers, State Historical Society of Wisconsin, Madison, Wis.

7. Cf. W. F. King to W. O. Atwater, July 9, 1881, Atwater Papers; Davenport Fisher to S. W. Johnson, S. W. Johnson Papers, Biochemistry Department, Connecticut Agricultural Experiment Station, New Haven, Conn.

8. Erwin Frink Smith to W. A. Henry, May 10, 1886, E. F. Smith Papers, American Philosophical Society, Philadelphia, Pa.; S. H. Gage to the President and Trustees of the University of North Carolina, August 25, 1881, S. H. Gage Papers, Collection of Regional History, Cornell University, Ithaca, N.Y.; Gage to

An experiment station position implied, moreover, a number of peculiar problems in addition to those normally faced by American scholars. Personality, for example, was an important consideration in hiring and firing, for all station workers were expected to be effective in contacts with farmers.[9] Another problem lay in the demand for regular publication; bulletins and annual reports had to make their appearance even if there were little or no original findings to fill them.[10] Research plans reflected economic needs, not the training and aspirations of staff members. If no entomologist were available, a chemist or botanist would simply have to be substituted to answer farmers' questions, prepare bulletins, and perhaps give spraying demonstrations. Vacations often had to be sacrificed to rural lecture tours. In some states, moreover, days were filled with the endless detail of regulatory work—not only fertilizer testing but the testing of seeds, horticultural varieties, and the like.

No problem was more exasperating to station scientists than the assumption that they should be responsible for answering any and all questions which might be addressed to them. Indeed, a number of directors were proud to list among their accomplishments the fact that their stations had become "general bureaus of information." This achievement seemed far less positive to staff members increasingly occupied in answering a bewilderingly varied correspondence. As one station scientist complained in 1881, "This is the *very busiest* time of the year for us. Farmers are busy making their plans for the coming year & they write to me on all sorts of subjects.

W. Trelease, September 3, and September 12, 1881, William Trelease Papers, Collection of Regional History, Cornell University; Louis F. McInnis to William L. Broun, April 27, 1889, Broun Papers, Department of Archives, Auburn University, Auburn, Ala.; W. B. Stark to J. K. Patterson, May 6, 1889, Patterson Papers, University of Kentucky Educational Archives, Lexington, Ky.

9. For example, Cornell felt some doubt in considering the promising young Alabaman Benjamin Duggar for an experiment station position because of some possible "southern peculiarity" which might interfere with his success in meeting New York farmers. George Atkinson to W. G. Farlow, April 15, 1896, Farlow Papers, Farlow Herbarium, Harvard University, Cambridge, Mass.

10. See F. B. Mumford to J. W. Connaway, April 25, 1905, University of Missouri College of Agriculture Papers, Columbia, Mo. "In its present character," an editorial complained in 1887, "the agricultural experiment station bulletin is a far less important document than it should be. The cause of this publication is ostensibly to report progress, practically to propitiate a constituency who unreasonably expect the most important problems settled as though by sleight of hand." *Agricultural Science*, I (March, 1887), 55.

And the trouble is every one has an equal right to an answer. I am a public man, and belong to the people & the people are relentless and exacting in their demands." [11]

With the exception of those stations created by states before the Hatch Act, the majority of stations were departments of the several land-grant colleges. In this relationship, still another ambiguity confronted the would-be scientist. Proximity to a university—though in some ways an intellectual stimulus—implied drawbacks as well.[12] The most important was the tendency of college and university administrators to exploit experiment station budgets. The most common form of such exploitation lay in charging the salaries of men whose duties were essentially teaching college science courses to the experiment station fund. However, some administrators, showing a great deal more imagination, charged insurance, a portion of the president's salary, students' laboratory supplies, and, in one case, even a carpet to the Hatch fund.[13] Professors of agriculture at land-grant colleges were often appointed directors of the new agricultural experiment stations—at no increase in salary or reduction in teaching responsibilities. And, although administrators might demand a profit

---

11. C. W. Dabney to Mrs. Dabney, Dabney Papers.

12. One problem lay in the endemic status conflicts between scientists with station appointments—often in applied fields—and those in regular university departments. At Wisconsin, for example, the College of Letters and Sciences sought at one point to limit appointments in the College of Agriculture to men whose work lay exclusively in applied science. Cf. W. D. Hoard to Charles Van Hise, October 12, 1908, and Van Hise to Hoard, October 14, 1908, President's Papers, University Archives, University of Wisconsin, Madison, Wis. A half-year later, Hoard complained again to President Van Hise of the "ill-concealed sneers of certain professors in the faculty of the College of Letters and Science toward the College of Agriculture." Van Hise replied weakly that jokes about a "cow university" were not meant seriously. Hoard to Van Hise, April 16, 1909, and Van Hise to Hoard, April 19, 1909, President's Papers. Yet available evidence seems to indicate that proximity between state university and college of agriculture—as at Cornell, Illinois, Wisconsin, and Minnesota—was on the whole more intellectually satisfying and effective than separation of the institutions.

13. See A. C. True to R. H. Miller, January 17, 1897, Records of the Office of Experiment Stations, Maryland File, National Archives, Washington, D.C. (hereafter cited as OES); True to H. R. Clark, April 5, 1895, OES, Oregon File. "The average agricultural college trustee," commented C. S. Plumb, a scientist with three years' experience under the Hatch Act, "believes that the government appropriation for experiment stations is a sort of windfall to the colleges, and is to be used to help these institutions." *Agricultural Science*, IV (April, 1890), 102.

from farm operations, they ordinarily applied this income to general college purposes. Not surprisingly, station scientists often opposed these practices which denied them time and resources for investigation.

Able and articulate experiment station scientists and administrators responded to these practices by seeking to manipulate an environment which seemed on every hand to compromise their autonomy. Idealism, as well as self-interest, at once prompted and legitimated such policies. Few experiment station scientists could accept the possibility that an irreconcilable conflict might indeed exist between the scientist's needs and those of his agricultural constituency, for few doubted that the conditions most appropriate for "high-grade" research were the optimum conditions for economic growth. A career in station work almost demanded such a point of view; the scientist's need to create a rapprochement between the demands of his own discipline and the requests of farm spokesmen made this formulation a logical and emotional necessity.

Some experiment station scientists did refuse to accept any version of this justifying doctrine, convinced that the incessant demands of the stations' agricultural constituency made any true scientific vocation impossible. An example is the experiences of students of Harvard mycologist W. G. Farlow, who, unable to secure "appropriate" university positions, became botanists and plant pathologists in experiment stations. The scorn of a man like Roland Thaxter for the Connecticut farmers who besieged him with demands for practical help was acid and unbending. "Bordeaux mixture is the vilest compound imaginable," Thaxter complained to his teacher Farlow, "but it would give me intense satisfaction to spray a select committee of Connecticut farmers with it till they couldn't see out of their eyes and the moss started from their backs, and then to soar away to lands where farmers are unknown and blight, blast, mildew, mould, rot, rust, scab, scald, smut, and spot are known by better-tasting names." Other station scientists, oriented toward a "secular" careerism, used misleading rhetoric and political manipulation to achieve an at least temporary success. It should be emphasized that most station members filled a varied spectrum between the genteel scorn of a Thaxter and the unalloyed boosterism of certain other station men.[14]

THE SCIENTIST-ENTREPRENEUR

The professionalization of station science entailed two principal social roles. One was that of the working scientist; the other, the more specialized role of the research entrepreneur—a role more important in the twentieth century, but evident in the formative years of the American experiment stations. These leaders, usually station directors, were forced to mediate between the world of science, on the one hand, and the social and economic realities of a particular state constituency, on the other. The best of these research entrepreneurs (who might or might not be productive scientists themselves) forged political alliances of mutual convenience with farm leaders and businessmen. At the same time, however, they had to maintain some commitment to and understanding of the world of science. That is, the successful research entrepreneur not only tailored a research policy to the needs of his constituency, but remained aware of professional values and realities; institutional success was dependent on the research entrepreneur's political skills as well as on his ability to recognize and hire men of talent and to provide them with an at least minimally adequate work environment.

Inevitably, a role so complex and demanding was pioneered successfully by only a handful of men. Also inevitably, client-oriented policies, which had brought success in the last decades of the nineteenth century, became increasingly compromising and irrelevant as the twentieth century progressed. In any case, the forging of this entrepreneurial role was still a necessary step in the development of the agricultural experiment station (as it was in the development of twentieth-century scientific institutions generally). Three successful research entrepreneurs were W. A. Henry of Wisconsin, Eugene Davenport of Illinois, and E. W. Hilgard of California; each sought

---

14. For dissenting views, see R. Thaxter to Farlow, April 14, 1890, Farlow Papers; cf. Thaxter to Farlow, March 4, and December 31, 1889, March 13, 1890; and W. C. Sturgis to Farlow, March 29, 1898, and June 14, 1900, *ibid.*

and achieved institutional security. The parallels are obvious and instructive.

At the University of Wisconsin, the first formal provision for agricultural experimentation had been made two decades before the passage of the Hatch Act in 1887, and an "experimental farm" had existed on the Madison campus since 1866. However, it was not until the arrival of William A. Henry, in 1880, that Wisconsin's agricultural program began to make progress.[15] A graduate of Cornell's demanding four-year course in agriculture, Henry had had no further advanced training; he was both in function and in his own perceptions a classic prototype of the research entrepreneur. Appointed as professor of agriculture, he realized immediately that it would be impossible to build institutional strength from the college's teaching program; students simply failed to attend the college. "The other line of work left open to us is experimentation," he explained to the Board of Regents. "By advancing agricultural science and getting our farmers interested in our work, we can justly hope, I hope, for more prosperous days in the future."[16]

Henry's attitude toward the success of the station and the College of Agriculture was based on three policy assumptions. The first was his recognition of the necessity for improving relations with the state's influential farmers combined with his conviction that this could be accomplished only through tailoring research to the farmers' needs and then convincing this skeptical constituency of the station's potential usefulness. Second, Henry realized that the ability to solve such problems could come only through the work of scientists with the best possible training. Third, he realized that the success of both college and station would depend on the ultimate success of Wisconsin's agricultural community in adjusting to a postgrain economy.

---

15. Board of Regents of the University of Wisconsin, *Annual Report for the Year Ending September 30, 1866,* p. 8. See also W. H. Glover, *Farm and College: The College of Agriculture of the University of Wisconsin: A History* (Madison, Wis., 1952); Vernon Carstensen, "The Genesis of an Agricultural Experiment Station," *Agricultural History,* XXXIV (1960), 13–20; Edward H. Beardsley, *Harry L. Russell and Agricultural Science in Wisconsin* (Madison, Wis., 1969).

16. "Report of the Professor in Charge of the Experimental Farm," Board of Regents of the University of Wisconsin, *Annual Report for the Fiscal Year Ending September 30, 1882,* p. 44.

Not surprisingly, Henry's most strenuous efforts were directed toward the conversion of the state's wealthier and more influential farmers. He insisted that all his men, no matter what their personal inclinations, play an active role in popularizing the station and the College of Agriculture. "I have a man," Henry explained whimsically, "I have been training for seven years, and though he groans when he sees a visitor coming and feels like running, he knows he has got to stay and take it." Henry and his staff explained again and again to visitors that an experiment station demanded an excellent chemical laboratory, that it was not a model farm and could not be expected to make a profit. "I have done lots of it," Henry reminisced:

> The first thing we are met with every day is this statement: "Well, professor, this farm will be satisfactory when you make it pay." . . . I knock that man right down and drag him off the farm. Someone says: "Professor, what does it cost you to make a pound of butter?" expecting me to say 15 or 18 cents, but I say, "Gentlemen, most of our butter this winter is costing us $5 a pound, I think." I knock them right down at once and then explain.[17]

In addition to such face-to-face contacts, Henry wrote widely for farm papers and lectured at Grange meetings and farmers' institutes throughout the state. He sought, in addition, to find ways of bringing farmers to Madison. Toward this end, Henry, a vigorous educational innovator, pioneered the use of practical winter short courses as a means of spreading the gospel of scientific agriculture and, at the same time, of building support for the university.

Interested in several areas of agricultural production, Henry chose to endorse dairy farming with particular vigor. By the mid-1880s, he had begun urging Wisconsin farmers to adopt dairying as a remunerative and stable response to a changing market.[18] Henry hired

---

17. Remarks of W. A. Henry, *Proceedings of the Second Annual Convention of the Association of American Agricultural Colleges and Experiment Stations January 1, 2, and 3, 1889,* OES Miscellaneous Bulletin No. 1 [Washington, D.C., 1889], p. 37.

18. It should be noted that the Wisconsin State Dairyman's Association had sought to encourage scientific study of their industry's problems as early as 1875. Glover, *Farm and College,* p. 81. See also Eric E. Lampard, *The Rise of the Dairy Industry in Wisconsin: A Study in Agricultural Change 1820–1920* (Madison, Wis., 1963), an excellent study of Wisconsin's dairy industry that clearly documents the role of the experiment station in its growth.

S. M. Babcock as station chemist in 1887 because of the Göttingen-trained New Yorker's demonstrated success in milk chemistry. With Henry's talent for choosing first-rate investigators and his understanding of how such intellectual resources might best be applied in economic contexts, by 1905 the Wisconsin station had contributed a number of innovations instrumental in the reshaping of traditional dairy practice: the proper use of silage, the cold-curing cheese process, and the Babcock test, for example.

As a successful research entrepreneur, Henry was blessed both with an understanding of the economic needs of his state's agricultural community and a measure of sympathy for the professional needs and attitudes of scientists. When hiring both Babcock and dairy bacteriologist H. L. Russell, who would be Henry's successor as college dean and station director, Henry was careful to emphasize the lightness of teaching duties and the opportunities for research which their positions entailed.[19] And, though Wisconsin's presidents were generally friendly to his work, Henry discouraged even small encroachments upon research funds for general university purposes. By judicious manipulation of agricultural politicians and editors and through an ability to produce tangible results, Henry and Russell had, by 1914, raised state appropriations for agricultural research to the then handsome figure of $75,000.[20]

This accumulation of institutional strength had certain negative concomitants. Perhaps most important was Henry's commitment to the need for "practicality" in research, a commitment justifiable perhaps in the 1880s but decreasingly fruitful in the first decade of the twentieth century.[21] ("As between the so-called theorist and the practical man," he wrote just after leaving his administrative duties

---

19. Henry to Babcock, August 31, and September 16, 1887, Babcock Papers.

20. In 1914–15, the University of Wisconsin College of Agriculture and Experiment Station expended $127,180 for research. (The Adams and Hatch appropriations totaled $30,000, the state funds came to $75,456, and farm sales provided another $21,724. In addition, the state legislature provided $10,000 for the publication of Experiment Station bulletins and circulars.) F. B. Morrison to J. H. Skinner, March 28, 1916, College of Agriculture Papers, University of Wisconsin Archives. See also Henry to Van Hise, February 23, 1907, President's Papers, University of Wisconsin Archives.

21. And practicality had, of course, to be judged not only by Henry, but by his evaluation of what farm leaders would be willing to deem practical.

in 1908, "I would take the practical man every time.")[22] Henry could be a harsh taskmaster for those scientists unwilling or unable to commit themselves to agricultural improvement at the cost of sacrificing an opportunity for achievement within their particular discipline.[23] Moreover, the heritage of justification in terms of popular approval could handicap even the most creative laboratory men. Biochemist E. V. McCollum, for example, a discoverer of Vitamin A, at first found it impossible to get financial support for the rat colony which he hoped to establish for nutrition studies; the Wisconsin legislature could hardly be expected to look with favor on tax monies being used to provide board and room for the farmer's bitter enemy.[24] Similarly, geneticists at Wisconsin and elsewhere anticipated opposition to plans for using small mammals such as rats, mice, and guinea pigs in breeding experiments; why, critics charged, should state and national funds pay for information on the breeding of pests? Ultimately even more compromising was the heritage of extension work created by Henry's policy of attracting farm support through service. These politically strategic contacts with farmers had by the turn of the century produced an exhausting load of correspondence and lecturing.

The central irony lay in Henry's very ability to gather state-wide support. The mechanisms which brought this success necessarily compromised the working scientist's autonomy. Yet, as contemporaries argued, no other approach would have been appropriate to the social and political realities of the 1880s and 1890s. Henry's tactics were, indeed, so successful that administrators in a number of other states found in them a model for their own policies.

No research entrepreneur was more assiduous in such efforts than Eugene Davenport, Henry's counterpart at the University of Illinois.

---

22. Henry to Truman Fullenwider, March 3, 1908, College of Agriculture Papers, University of Wisconsin Archives.

23. Cf. Trelease to W. G. Farlow, June 22, 1882, Farlow Papers; H. P. Armsby to E. H. Jenkins, July 20, 1886, Director's Papers, Connecticut Agricultural Experiment Station, New Haven, Conn.; and Babcock to Mrs. Babcock, May 23, and September 5, 1897, September 7, 1899, Babcock Papers.

24. Cf. L. B. Mendel to T. B. Osborne, July 26, 1912, Biochemistry Department Records, Connecticut Agricultural Experiment Station, New Haven, Conn.

Though perhaps less successful than Henry in promoting real contributions to agricultural productivity—possibly because he was less skillful in his choice of staff—Eugene Davenport was even more successful in garnering support for his research program. Agricultural education and research had borne comparatively little fruit at Illinois before Davenport was hired in 1895 as dean of the College of Agriculture and director of the Illinois Experiment Station.[25] With feeble backing from a president uninterested in the college and the station, Davenport was forced to turn for support to the state's agricultural producers and agriculture-related industries. A man of immense political sensitivity, Davenport was quick to cultivate the leaders of the Illinois specialized producers' associations and gained legislative strength by having these representatives of the state's commercial agriculture lobby for categorical appropriations— for research in soils, crops, livestock, and even floriculture. "There is no question," Davenport explained again and again, "but that money devoted to investigation pays, and pays immediately." "In the final round-up before the legislature," he confided to an aspiring Iowa administrator, "nothing counts so much as a delegation of actual farmers representing organizations. Their influence does not depend upon their numbers; while anything like a popular movement, when organized, requires a good deal of work to arouse it in the first place, and is liable at any moment to subside." Ultimately, of course, Davenport realized, he would have to seek research funds without hemming them in with specific requirements.[26]

Ironically, however, his success in building support for the college and station was to make this desired autonomy difficult to achieve. Even more than Henry, Davenport became the prisoner of his own political acumen. The state legislature not only acceded to the re-

---

25. See College of Agriculture Letterbooks, 1888–1911; Experiment Station Letterbooks, 1901–4; and Agronomy Letterbooks, 1899–1919, University of Illinois College of Agriculture, Urbana, Ill.

26. Davenport, *The Work and Needs of the Agricultural College and Experiment Station of the University of Illinois, February, 1903,* (n.p., n.d.), p. 12; Davenport to W. J. Kennedy, February 1, 1906, Dean's Letterbooks, Illinois College of Agriculture; Davenport to C. F. Mills, October 19, 1900, Dean's Letterbooks.

search requests of Davenport's allies, but it institutionalized the relationship by creating an advisory committee of prominent agriculturists to oversee the several lines of investigation the state had opted to support. (The Illinois Grain Dealers Association and Corn Growers' Association, for example, jointly appointed the committee on crops; the Illinois State Horticultural Society, the committee on orchards; the Dairyman's Association, the committee on dairying; and so forth.) Enmeshed in this intimate relationship, Davenport found it difficult to consistently capture the initiative in research policy-making. Other problems followed. Farm leaders directly influenced hiring and firing, while large appropriations brought pressure for immediate results. In 1911, for example, Davenport was forced by the state's dairy association to ask for the resignation of the head of his dairy husbandry department.[27] Such close relations between experiment station staff and their clients also maximized the likelihood of nonprofessional conduct; incidents of premature disclosure of work in progress accompanied by grandiose claims, as well as staff members' involvement in indiscreet business enterprises, troubled Davenport's directorship.

Yet, by the First World War, he could point to a record of outstanding growth for Illinois' College of Agriculture and Experiment Station. In terms of staff, research support, and number of students, the college and station had become firmly established. This is demonstrated vividly in Davenport's ability to bypass Presidents Edmund J. James and Andrew S. Draper (neither of whom was an enthusiastic advocate of agricultural education and research) in gaining access to the legislature.[28] Davenport was ultimately able to mobilize a network of state-wide influence to gain support for general university needs. By 1910, the university's orphan child had become at once a leading advocate of and argument for legislative largesse.

---

27. Davenport to E. J. James, April 11, 1911, Dean's Letterbooks.

28. As Davenport explained to a contemporary, "A man desiring to develop a department of economics or history, Greek or philosophy, is practically confined to internal conditions for the achievement of his ambition; fortunately it is not so with agriculture. The state of South Dakota, not the campus at Brookings, is your field. . . . If you can make your department indispensable to South Dakota, there is nothing you cannot have." Davenport to A. N. [Hume], October 9, 1918, Davenport Papers.

Of the three brief examples, E. W. Hilgard of California is perhaps the most complex. Hilgard was both scientist and research entrepreneur, a geologist of international reputation, and at the same time a gifted manipulator of public opinion.

When he arrived in Berkeley as professor of agriculture in 1875, Hilgard found a campus beset with problems: the Grange was actively hostile to an institution in which agriculture was clearly held in small regard;[29] the Berkeley agriculture college was unsuccessful; and Hilgard, like Henry and Davenport, turned to research as justification for continued legislative support. Hilgard immediately set out to placate the Grange through personal contacts and by undertaking work relevant to the needs of California agriculture, especially in viticulture—most notable was his campaign against the phylloxera disease—and, closer to his original interest and training, the study of California's varied soils. Central to these plans was an experiment station. (Even before arriving in Berkeley, Hilgard had sought to popularize the need for American agricultural experiment stations on the German model.) Within a few years of his arrival, Hilgard could boast of such a station, in function, if not in name. By the mid-1880s, Hilgard had soothed the opposition and built a respectable base of support for his work.[30]

The logic of Hilgard's position implied the almost exclusive emphasis upon immediate economic goals; by the 1890s he was forced to defend his service-oriented policy against the frequent disapproval of the Office of Experiment Stations and other advocates of more fundamental station research. With California's novel soil and climatic conditions, Hilgard argued, such policies were the only appropriate response. These were the arguments Hilgard employed in his justification of his policy of encouraging branch stations throughout the state to the Office of Experiment Stations, which vigorously opposed this plan. Hilgard was, moreover, forced

---

29. See E. W. Hilgard Papers; and Mary L. Mayfield, "The University of California Experiment Station, 1858–1900" (Master's thesis, University of California, Davis, 1966).

30. Years later, Hilgard was happy to submit a claim for priority in establishing America's first experiment station. Hilgard to A. C. True, January 7, 1892, OES, California File. See also Amos Adams to Hilgard, October 6, 1879, Hilgard Papers.

to embrace, at least rhetorically, the working farmer and rancher. Despite his academic credentials, Hilgard assured farm audiences, he had had personal experience as a farm owner and understood their problems. He was not a professor who "fights shy" but one "who has been there and knows how it is himself." As a result of his responsiveness, however, Hilgard soon found that the demands made upon his time had become oppressive. "There is no rest here for anyone, wicked or otherwise," Hilgard complained to F. A. P. Barnard in 1886, "least of all for a man who, like myself, is in a position which authorizes everyone from the shock-haired and hay-seed-bestrewn granger to the justices of the supreme court to ply me with questions on their private business." His routine correspondence alone, he complained, would "do credit to a wholesale business house."[31] However, as was the case in Wisconsin and Illinois, such contacts were unavoidable if the work was to find adequate support.

Like Davenport of Illinois, moreover, Hilgard had to contend with an unsympathetic president and several uncooperative regents. As dean of the College of Agriculture, he thus had to wage a bitter biennial campaign for appropriations. Generally successful in his lobbying efforts, Hilgard looked forward in the early 1880s to the availability of permanent federal support for his state's experiment station; he worked vigorously for passage of the Hatch Act. "I am getting heartily tired of my biennial fight for existence," he explained to a friend.[32] Even after the passage of the bill, he continued to lobby in Sacramento for the categorical appropriations which helped ensure the continued growth of his college and station. Hilgard was justifiably proud of his achievements: "I have built up the whole," he wrote to President George Atherton of Pennsylvania State University, "by an aggressive policy in the face of many discouragements, even the regents tolerating it at first only as a sop thrown to the Cerberus of the Grange. But the work has acquired

---

31. Hilgard to F. A. P. Barnard, January 1, 1886, Hilgard Papers; Hilgard to A. L. Bancroft, March 25, 1886, *ibid.*; "Address to Agricultural and Horticultural Society," *Los Angeles Daily Republican*, October 7, 1877.
32. Hilgard to Harvey Wiley, July 29, 1885, Hilgard Papers.

an impetus that renders it unsafe for legislators or regents to ignore it."[33]

THE RESEARCH SCIENTIST AND HIS CONSTITUENCY

Differences in personality and local economic conditions implied minor tactical differences, but, in all three cases, the research entrepreneurs' strategies were the same: the creation of institutional strength through alliance with members of the business, agricultural, and political communities. In terms of the scientist's immediate work environment, these relationships could, as we have seen, become a mixed blessing. Even those nonscientists most sympathetic to the stations and most euphoric in their hopes for scientific agriculture often entertained quite narrow views of the tasks appropriate to the experiment station scientist.

Scientists reacted in several ways to such constraints. The principal response was to work, as Henry, Davenport, and Hilgard did, for power and autonomy within existing sources of economic and political influence. Another was to appeal to competing sources of power and reassurance. One such source was the United States Department of Agriculture's Office of Experiment Stations. (These recourses were particularly important for scientists in states where deans and directors were more attentive to the demands of local clients than to the desires of their own staff members.) A third reaction was to create new applied science disciplines with norms more appropriate to the demands of the experiment station context than those of older pure science fields. And, finally, agricultural scientists in almost every discipline reassured themselves and attempted to "educate" their clients by endorsing a rhetorical stance which emphasized the need for pure research if applied science were to prove ultimately fruitful.

The Office of Experiment Stations played a particularly strategic role in attempts by station scientists to shape their professional lives. Under the control of men sympathetic to the scientist's point of

---

33. Hilgard to G. W. Atherton, December 6, 1884, Hilgard Papers.

view, the OES administration was consistently alert to the dangers of local obstructionism. In the 1890s, the Office gained increased powers of audit and inspection, powers which it used to discourage the exploitation of Hatch funds.[34] (Such exploitation meant, in day-to-day terms, either the scientist's assignment to wearying teaching and extension work or the diversion of funds which might have benefited research.) Moreover, functioning as a semiformal employment agency for workers in the agricultural and biological sciences, the Office was able to seek jobs for innocent victims of local power conflicts, and find better places for the competent. In the early years of the twentieth century, the elite among experiment station leaders united with the OES to lobby for passage of the Adams Act (1906), which doubled the funds available for the state stations, but with the important and innovative proviso that the money be spent for "original investigation." The "original investigation" provision of the Adams Act provided greatly increased administrative leverage for the scientifically-oriented administrators who controlled the OES. One wrote at the time:

> It is an opportunity which has never been presented before the world; and if embraced our stations may easily assume leadership in agricultural investigation. . . . It will take a good deal of patience and arguing for lots of people think they are doing investigating work if they are conducting some variety tests or testing South Carolina rock against ground bone for potatoes.[35]

Four years later the Smith-Lever Act, also supported by experiment station leaders and the OES, began to provide funds for extension work and thus removed another burden from limited station funds and staff.

---

34. Although the power to disallow claims was rarely invoked, the OES used the threat constantly and made annual inspection trips to the state stations. See Milton Conover, *The Office of Experiment Stations: Its History, Activities, and Organization,* Institute for Government Research, Service Monographs of the United States Government No. 32 (Baltimore, Md., 1924); Leonard D. White, *The Republican Era, 1869–1901: A Study in Administrative History* (New York, 1958), pp. 247–52.

35. See Charles E. Rosenberg, "The Adams Act: Politics and the Cause of Scientific Research," *Agricultural History,* XXXVII (1964), 1–10; E. W. Allen to J. B. Lindsey, May 8, 1906, OES, Massachusetts File.

Scientists also appealed to their disciplinary peers for moral support and endorsement and to such values as academic freedom and professional competence against local political or administrative pressures. Scientifically-oriented station men also charged that administrators who catered blindly to their constituents were performing a disservice to American agriculture as well as to freedom of inquiry. This view became pervasive by the turn of the century. Most American agricultural scientists adopted a rhetorical stance which justified pure science as a necessity if applied science was to function properly; only upon a base of high-quality research could ultimate economic gains be predicated. This was a necessary stance for the majority of station scientists, who could neither consider themselves mere implements of their agricultural constituency nor reject the structure of compromise exacted by this clientele.

Indeed, this argument and its corollary that all insights gained into the laws of nature would be transformed through some serendipitous process into immediate practical benefits were so frequently reiterated that they became sources of unmet hopes and unreal expectations. For instance, in the first dozen years of this century the insights of Mendel and DeVries seemed not only to imply the creation of a new and promising discipline but to guarantee instant practical results for the farmer. (One station writer noted slyly at the time that: "Members of the legislature who have the dispensing of funds will certainly find more comfort in the theory of DeVries than in that of Darwin.")[36] These ingenuous hopes were soon blighted; Mendel added little or nothing to the technical resources of the skilled empiric breeder. Geneticist Raymond Pearl noted that all that the new genetics had really done before World War I was to help the breeder interpret his customary techniques. He explained,

This may seem too mild a statement of the potential value of genetic science to the animal breeder. It undeniably does lack the grandeur of the vision sometimes opened out by the extension

36. N. E. Hansen, "Plant Breeding," *Proceedings of the Eighteenth Annual Convention of the Association of American Agricultural Colleges and Experiment Stations, November 1–3, 1904,* OES Bulletin No. 153 (Washington, D.C., 1905), p. 119.

lecturer in his zeal to inspire the farmers to do better things, and at the same time pave the way for increased appropriations for his institution.[37]

The rhetorical emphasis upon the necessary, mutually beneficial relationship between pure science and applied science began to change at the end of the nineteenth century. The debate over this traditional emphasis grew into a conflict between an older generation of scientist-administrators strongly committed to the legitimacy of the demands made by their agricultural clients and younger— often European- or graduate-school-trained—laboratory men. These younger scientists tended to assume much higher ground in discrediting policies which allowed the perceived needs of a farm constituency to define research programs. As state support for stations increased in the early years of the century, and as the Adams fund became available after 1906, American experiment stations turned increasingly to the nation's doctoral programs for staff members. Agricultural colleges produced almost no doctorates, and station directors had—in some cases unwillingly, but with the consistent encouragement of the OES—to turn to basic science departments for well-trained men.[38] As their numbers increased, criticism of the older generation of politically sensitive research entrepreneurs grew sharper and more frequent.

Reformers were willing to concede that accommodation was necessary in the pioneer years of the experiment station movement and was appropriate both to political realities and to the stations' limited technical capacities. However, by 1900, articulate critics submitted, American agricultural research should have outgrown such immature relationships; it was no longer necessary to dramatize the possible relevance of station science to the farmer's economic

---

37. Pearl, *Modes of Research in Genetics* (New York, 1915), p. 169.

38. For example, within a few years of T. H. Morgan's first drosophila publications, Office of Experiment Station administrators were urging state directors to turn to the Morgan group at Columbia in their search for men to conduct breeding programs. E. W. Allen to H. W. Mumford, January 2, 1912, OES, Illinois File. A number of even the more alert and responsive members of the older generation of scientist-administrators felt, however, real qualms in regard to hiring scientists with no specific farm training or orientation. Cf. L. H. Bailey, "Training for Experimenters," *Agricultural Science*, V (1891), 214–15; H. J. Waters to E. W. Allen, March 24, 1902, OES, Missouri File.

needs. American scientists should emulate their German counterparts and educate the public and administrators to the need for higher standards of research if substantial increases in productivity were to be attained. How else, the argument inevitably continued, could real economic progress take place? The dissemination and adaptation of existing knowledge were intrinsically limited activities; the people must be rescued from their ignorance and made to understand the nature of true scientific research.[39]

THE STATIONS AND THE GROWTH OF SCIENCE

It would be easy enough to emphasize ways in which the experiment station context and traditions played a negative role in the development of the sciences in America. A tradition of client-centered research led to much trivial and redundant work. The need to achieve institutional and individual autonomy did result in a habit of compromise—a habit easily justified by the democratic rhetoric of service as ultimate goal.

Despite such constraints, it can be argued that the agricultural experiment station played, in sum, a positive role in the development of the scientific disciplines in the United States. Though the relationship between agricultural scientists and their economic context was ambivalent, this period served as a formative stage in the gradual shaping of a research context increasingly hospitable to more fundamental work.[40]

In specifying components of this complex relationship, one can begin with a positive generalization. The stations played a substantial role in the careers of individual scientists and in the formative period of a number of the biological sciences in America—particularly bacteriology, biochemistry, and genetics. The reasons are obvious. In years of austere academic budgets, almost any posi-

---

39. See, for example, W. H. Jordan, "Function and Efficiency of the Agricultural College," *Science,* XXXIV (1911), 780–81.

40. Analogous economic pressures helped as the twentieth century progressed to support a much closer and more organic relationship between basic science and agricultural technology. The comparatively rapid development of the scientific disciplines in the land-grant colleges and experiment stations surprised even contemporaries in the early years of the century.

tion was a good one. A job in which research and publication were formal aspects of one's duties seemed particularly attractive; though often disappointed, youthful scientists were again and again attracted to experiment station work. The better state stations and colleges of agriculture offered greater freedom than the United States Department of Agriculture,[41] while research positions in industry were still scarce indeed. Despite the narrowness and frustrations of the experiment station's applied science environment, station opportunity might be structured so as to make its expectations at least partially consistent with the aspirations and demands of the scientist's own discipline.

The most important direct relationship between the experiment stations and the development of the scientific disciplines in America was the strength the stations gave to the state universities. The strategies for the accumulation of political influence adopted by college and station interests helped bring about a more secure level of support for many state universities; the relationship between college and station—generally an ambivalent and ambiguous marriage —was essentially constructive. Examples of successful technological innovation made investment in the production and dissemination of such innovations attractive and at the same time created a group of potential lobbyists in farmers and businessmen who had themselves benefited from station work. And, of course, whatever role the stations played in increasing agricultural productivity ultimately increased the resources potentially available for expenditure on public purposes.

Indeed, although it may be argued that the relatively crude level of pre-World War I agricultural technology served to limit the stations' capacity for actually increasing productivity, the role of the experiment station in strengthening the state university became increasingly important as the twentieth century progressed and a more intricate and fruitful relationship between basic and applied science became both necessary and possible. For example, at Wisconsin

---

41. See J. W. Coulter to E. F. Smith, July 14, 1888, Smith Papers; W. J. Spillman to A. F. Blakeslee, November 1, 1911, Blakeslee Papers, American Philosophical Society; H. J. Conn to K. F. Kellerman, May 1, 1911, Conn Papers, Collection of Regional History, Cornell University.

extremely strong departments of bacteriology, biochemistry, and genetics all grew up within the College of Agriculture and experiment station. The increasing success of the college of agriculture as an educational institution also played an important role in economic growth, helping create common values in community leaders and thus conditions in which receptivity toward innovation and respect for the role of science were maximized. The experiment station, with its professionally oriented staff of disseminators, adaptors, and innovators, existed to provide a nexus in which such relationships could be created and exploited.[42]

THE STATIONS AND ECONOMIC GROWTH

At least as important, perhaps, as the direct role played by the stations in the development of American science was the effect they exerted on economic growth and thus, indirectly, on science as well. Obviously the distinction between the role of the stations in promoting economic growth and in developing the scientific disciplines is arbitrary at best. From the perspective of the 1970s this arbitrariness is particularly obvious; for connections between basic science, technology, and the economy have become increasingly close and productive. Despite the difficulty of evaluating specific inputs and relationships in a system so complex and interdependent, it is possible to discern some of the ways in which the existence of the stations and the consequent movement toward professionalization among station scientists served as a component of economic growth.

The gradual professionalization of the experiment station scientist constituted a specific input in the system of agricultural productivity. In other words, the agricultural scientist's perception of an appropriate role and his efforts to narrow the gap between this perceived role and his originally unsatisfactory circumstances altered his behavior and thus served as an agent of economic change. This influence was

---

42. A number of tangible and permanently significant results of pre-World War I experiment station work, hybrid corn or the discovery of Vitamin A, for example, grew out of this complex background. It is difficult to discuss briefly their relationship to economic context, on one hand, and the community of pure science, on the other.

exerted in two general ways: in innovation and adoption and in the provision of mechanisms for the dissemination of resultant knowledge.

Whether the experiment station scientist defined his role in terms of applied science or some more traditional discipline, conformity to a professional role implied publication, mastery of a literature that was growing in both quantity and complexity, and, more fundamentally still, commitment to his fellow scientists' acceptance as an ultimate measure of achievement. Such acceptance was defined in terms of increasingly particular knowledge and increasingly detailed publications. Even the creation of new applied science disciplines such as economic entomology, horticulture, or poultry husbandry were shaped by these general criteria as well as others more discipline-specific.[43] In addition to values shared by scientists in general, agricultural scientists tended to accept service as an absolute value. In the era before the Smith-Lever Act, this ethic of service helped to create an extension program, in function if not in name, to bring and adapt agricultural innovation to an industry normally resistant to change.

The political needs of the station scientists guaranteed that the educated, adequately capitalized farmer would be their natural ally in the achievement of power. Indeed, the larger the scale of an enterprise, the more useful was the work of the experiment station scientists. Innovation and adoption implied both capital and appropriate values. And these values—particularly education—implied, in turn, a degree of prosperity. "I am sorry," W. A. Henry wrote in 1898, "to say that we have not accomplished much with the lower stratum of Wisconsin farmers, but the more intelligent ones who are leaders of their community and who help others by example, are for the most part strongly with us."[44] In the areas of inspection and control, the less-educated general farmer was often inhospitable to "book-farming" and to agricultural colleges, which seemed remote

---

43. See Charles E. Bessey, "A Duty Which We Owe to Science," *Proceedings of the Eighth Annual Meetings of the Society for the Promotion of Agricultural Science* (Columbus, Ohio, 1887), p. 28.

44. Henry to A. C. True, July 9, 1898, OES, Wisconsin File. The phrase "intelligent farmer" is a central one in the social and emotional shorthand of nineteenth-century Americans.

from his needs and those of his children. Similarly, the small nurseryman or fertilizer manufacturer frequently objected to station regulation; the larger producers often welcomed their attentions. When, for example, small nurserymen in North Carolina complained of inspection fees, the experiment station council decided that "it seems better to insist on their still paying even if there are many forced out of business. There is more danger by far from the small nurseries than the larger ones." Even in the earliest years of agitation for the creation of experiment stations and fertilizer inspection, large manufacturers of chemical fertilizers were often in favor of such programs.[45] The experiment stations served with some efficacy to strengthen, rationalize, and perhaps exacerbate trends already apparent within American agriculture and the broader society.

Logically enough, the stations tended to be most successful in states that developed a specialized agriculture oriented to urban markets and in states where respect for science and education as instruments for economic progress and moral improvement was most apparent. With a constituency eager to rationalize their methods and capable of being convinced by station arguments and demonstrations, conditions for the interaction of the needs of station scientists and the parallel development of agricultural productivity were maximized. By 1914 the effectiveness of station efforts in increasing productivity in particular states and particular industries seems clear enough; dairying was probably the most striking example.

Not only in Wisconsin, but in New York, Connecticut, Minnesota, Illinois, Iowa, Vermont, and New Hampshire, as early as the 1880s station directors actively promoted dairying. Working at first with a handful of large farmers, farm editors, and farm business leaders, some directors were able to create in the next decade a vigorous interest in this stable, year-round enterprise. A number of the more alert directors were sensitive to the economic contours of the budding dairy industry and thus were aware of production and

---

45. "Minutes of Station Council," November 1, 1897, Experiment Station Records, University Archives, North Carolina State University, Raleigh, N.C. Cf. B. M. Rhodes to Evan Pugh, April 16, and June 23, 1862, Pugh Papers, Penn State Collection, Pennsylvania State University, University Park, Pa.

marketing problems that could be solved by technological means. Thus, at least five stations were at work in the late 1880s on a practical butter-fat test for commercial creameries. S. M. Babcock, who had been hired by Wisconsin because of his skills in milk chemistry, perfected such a test. The test was described in 1890, and other stations acted quickly to popularize it among their constituencies. They not only published special bulletins, but also organized winter short courses in dairying with instructions in the Babcock test as a key inducement. The economic consequences of the Babcock test extended beyond the provision of a rational method for paying producers. It also served as a criterion for culling nonproductive animals from dairy herds and thus led to a general upgrading of dairy stock (including the more general adoption of the Holstein-Friesian).[46] Finally, of course, the test provided a public-relations argument for greater support of agricultural research and state universities generally. (Dean Davenport, for example, told the chairman of his state's committee on appropriations that the Babcock test alone had been worth more to the state of Wisconsin than the cost of running the entire university throughout its history.)[47] The Babcock test was only one among a series of technological innovations which helped to reshape the American dairy industry between 1890 and 1914. It must be recalled, moreover, that the existence of such technological innovations not only had a discrete effect on productivity but served as well to stimulate the industry by changing the image of dairying and attracting general farmers to this area of specialization—a special concern of station publicists.

It is, of course, impossible to make any categorical evaluation of the efficacy and wisdom of the investment of intellectual and capital resources in agricultural experiment stations by state and national governments before 1914. This is hardly a meaningful question. The problem was not formulated in these terms; it is hard to imagine how the historical circumstances which gave birth to the Ameri-

---

46. See Lampard, *Rise of Dairy Industry, passim;* Glover, *Farm and College;* and W. W. Cooke, "The Experiment Stations and Dairying," *Agricultural Science,* VI (January, 1892), 35–40.

47. Davenport to George H. Rankin, February 15, 1901, Dean's Letterbooks. The ultimately ambiguous quality of such dollars-and-cents arguments is apparent.

can experiment station movement would have allowed any alternative method of subsidizing agricultural research. Nor is it likely that these resources would have been diverted to other means of subsidizing pure or applied science. Neither individual units of production —the farmers—nor agriculture-related business would have considered such allocation of funds in the late nineteenth century. Only the political power of agriculture and the primacy of both agriculture and science in the American ethos allowed the stations to be created.

Interest group politics created the American experiment stations; the men who staffed these new institutions were motivated by their own values and institutional needs and in turn became an interest group, forging pragmatic alliances and formulating a research policy based on the general interest in the growth of productivity through the rational application of technology. The fragmented quality of American institutional life meant that the values of professionalization would—shaped into concrete institutional forms—establish themselves within this shifting intellectual and institutional landscape.[48] The station scientist's increasing commitment to professional norms was power, power to determine a pattern of behavior which stimulated economic development and the growth of both pure science and agricultural technology.

---

48. This evaluation of the place of American agricultural science and scientists in the nation's social and institutional structure is in marked contrast to David Joravsky's suggestive interpretation of Lysenkoian genetics in terms of Soviet realities. See "Cracked Wheat," review of *The Rise and Fall of T. D. Lysenko* by Zhores A. Medvedev, trans. Michael Lerner, *New York Review of Books*, XIV (January 29, 1970), 48–52.

Mirror-Image Twins:
The Communities of Science
and Technology

American technology went through a scientific revolution in the nineteenth century. Technological knowledge was uprooted from its matrix in centuries-old craft traditions and grafted onto science. The technological community, which in 1800 had been a craft affair little changed since the Middle Ages, was reconstructed as a mirror-image twin of the scientific community. The artisan was replaced in the vanguard of technological progress by a new breed. In place of oral traditions passed from master to apprentice, the new technologist substituted a college education, a professional organization, and a technical literature patterned on that of science. Equivalents were created in technology for the experimental and theoretical branches of science. As a result, by the end of the nineteenth century technological problems could be treated as scientific ones; traditional methods and cut-and-try empiricism could be supplemented by powerful tools borrowed from science. This change was most marked in the physical sciences and in civil, mechanical, and electrical engineering, the subjects of this chapter. But similar changes were taking place at the same time in the relations of chemistry, biology, geology, and other sciences to their corresponding technologies. The result might be termed "the scientific revolution in technology."

The significance, indeed the very existence, of the scientific revolution in technology has been obscured by a commonly accepted model of the relationship between science and technology. In essence, this theory holds that scientists create new knowledge, which technologists then apply. In 1831 Jacob Bigelow, articulating a common notion, asserted:

> Our arts have been the arts of science, built up from an acquaintance with principles, and with the relations of cause and effect . . . we have acquired a dominion over the physical and moral world, which nothing but the aid of philosophy could have enabled us to establish. . . . The labor of a hundred artificers is now performed by the operations of a single machine. We traverse the ocean in security, because the arts have furnished us a more unfailing guide than the stars. We accomplish what the ancients only dreamt of in their fables; we ascend above the clouds, and penetrate into the abysses of the ocean.

He concluded that "the application of philosophy to the arts may be said to have made the world what it is at the present day."[1] But when Bigelow enumerated the specific instances in which "philosophy" (science) had transformed technology, he noted such inventions as gunpowder, glass, the magnetic compass, the printing press, the clock, the cotton gin, the steam engine, and textile machinery. Yet in none of these cases is the influence of science on technology obvious; certainly in none is their relationship explained satisfactorily by the common model.

This view of the relationship between science and technology has continued into the twentieth century and has been voiced by Vannevar Bush and other architects of America's recent science policies. Bush held that basic research, albeit undertaken without consideration of practical ends, generates knowledge of nature's laws which provides the means of technological progress. He maintained that

> Basic research leads to new knowledge. It provides scientific capital. It creates the fund from which the practical applications of knowledge must be drawn. New products and new processes do not appear full-grown. They are founded on new principles and new conceptions, which in turn are painstakingly developed by research in the purest realms of science.

---

1. Jacob Bigelow, *Elements of Technology* (Boston, 1831), p. 4.

Bush concluded that "today, it is truer than ever that basic research is the pacemaker of technological progress."[2]

Inspired by Bush's model of the relation of science to technology, the Department of Defense, from 1945 to 1966, invested about ten billion dollars in scientific research, of which approximately one-quarter went for basic or undirected research. A growing skepticism concerning the technological value of this enormous expenditure caused the department to undertake an investigation, Project Hindsight. This study took eight years and consumed some 40 man-years of time on the part of thirteen teams of scientists and engineers who analyzed the key contributions which had made possible the development of the twenty weapons systems that constituted, in large part, the core of the nation's defense arsenal. Some 700 key contributions, or "events," were isolated. They were classified as being either technological or scientific. Of the latter, they were further subdivided into basic and applied science "events."[3]

The preliminary results of Project Hindsight, which were released in November, 1966, came as something of a bombshell to the scientific community. Of all "events," 91 per cent were technological, only 9 per cent were classed as science. Within the latter category 8.7 per cent were applied science; only 0.3 per cent, or two "events," were due to basic or undirected science.[4] Predictably, the publication of these results produced a spate of indignant letters to the editors of *Science*.[5] Many of these missed the point. The investigators had not sought to show that science has no influence on technology. What they did demonstrate was that the immediate,

2. Vannevar Bush, *Endless Horizons* (Washington, D.C., 1946), pp. 52–53. See also John R. Steelman, *Science and Public Policy* (Washington, D.C., 1947), pp. 4–5.

3. Chalmers W. Sherwin and Raymond S. Isenson, "Project Hindsight," *Science*, CLVI (June 23, 1967), 1571–77.

4. D. S. Greenberg, "'Hindsight': DOD Study Examines Return on Investment in Research," *ibid.*, CLIV (November 18, 1966), 872–73. Philip H. Abelson, "Project Hindsight," *ibid.*, CLIV (December 2, 1966), 1123.

5. See the collection of letters in "How Perceptive Is Hindsight?," *ibid.*, CLV (January 27, 1967), 397–98. See also Helen L. Hayes, "Project Hindsight: Basic Research," *ibid.*, CLIV (December 23, 1966), 1504; Allen M. Lencheck, "Project Foresight," *ibid.*, CLV (January 13, 1967), 150; Lee Leiserson, "Project Hindsight," *ibid.*, CLVII (September 29, 1967), 1512; and Robert M. Lukes, "Masquerade of Undirected Research," *ibid.*, CLIX (January 5, 1968), 34.

direct influence has been small; they showed that the traditional model of science-technology relations is in need of revision. To correct the misunderstanding of Project Hindsight, a subsequent study, TRACES, demonstrated the dependence of five recent innovations on prior scientific work. The question, therefore, is not whether science has influenced technology, but rather the precise nature of the interaction.[6]

The results of Project Hindsight are surprising only if one assumes the validity of the received model of science-technology relationships. This model is not so much false as misleading. It assumes that science and technology represent different functions performed by the same community. But a fundamental fact is that they constitute different communities, each with its own goals and systems of values. They are, of course, similar in that both deal with matter and energy. But these similarities should not be overstated. Each community has its own social controls—such as its reward system—which tend to focus the work of each group on its own needs. These needs determine not only the objects of concern, but the "language" in which they are discussed. These needs may overlap; but it would be surprising if this occurred very frequently. The expectation that science would beget more science and that technology would lead to further technology is supported for the twentieth century in the findings of Project Hindsight.

The difficulties of the traditional model may be illustrated by the relationship, or lack of one, between Newtonian mechanics and the "golden age" of mechanical invention in America in the nineteenth century. An enthusiastic group of scientists, technologists, and reformers in the United States, as in Europe, attempted to foster the application of science to technology. Among them was James Renwick, professor of natural philosophy and experimental chemistry at Columbia College. He wrote two books that were intended to bridge the gap between art and science. *The Elements of Mechanics*, published in 1832, was a conventional exposition of the science of

---

6. Illinois Institute of Technology Research Institute, *Technology in Retrospect and Critical Events in Science,* 2 vols. ([Chicago], 1968), I, 22.

mechanics. In it, Renwick followed a well-trodden path in treating systems in equilibrium by the principle of virtual velocities.[7] His *Applications of the Science of Mechanics to Practical Purposes,* published in 1842, surveyed the field of mechanical technology, including prime movers, clocks, and various types of machinery.[8] Despite Renwick's earnest efforts, the principles of the first book did not carry over to the second to any significant degree.

A mechanic interested in designing a water wheel would have found the methods and principles of the first book of little value, even if he was able to understand them, but he would have found valuable assistance in Renwick's second book. John Smeaton, the eighteenth-century British engineer, used the experimental methods of science to derive a set of "maxims" or design principles for this type of prime mover. Two examples are:

> In a given undershot wheel, if the quantity of water expended be given, the useful effect is as the square of the velocity,

and

> In a given undershot wheel, if the aperture whence the water flows be given, the effect is as the cube of the velocity.[9]

Neither could be classed as laws of nature; they were lawlike statements about man-made devices. They were not logical deductions from the science of mechanics; they constituted the germ of a new technological science.

Far from constituting a unity, Renwick's two books pointed to two quite different lines of technological development. As Renwick suggested in his first book, technology might build directly on the foundations of science. The science of mechanics could be extended to create new, technologically oriented sciences such as the strength

---

7. James Renwick, *The Elements of Mechanics* (Philadelphia, Pa., 1832), p. viii.

8. James Renwick, *Applications of the Science of Mechanics to Practical Purposes* (New York, 1842).

9. *Ibid.,* pp. 49–50; John Smeaton, "An Experimental Enquiry Concerning the Natural Powers of Water and Wind to Turn Mills, and Other Machines, Depending on a Circular Motion," *Philosophical Transactions,* LI (1759–60), 118–20.

of materials and hydraulics. Or, following Smeaton, technologists might borrow the methods of science to found new sciences built on existing craft practices.

To some extent, inventors helped to develop technological sciences. Oliver Evans attempted to apply scientific methods to technology in his *The Young Mill-Wright and Miller's Guide,* published in 1795. Evans surveyed the principles of mechanics and was able to derive useful design principles directly from them. But his chief reliance was on the application of scientific methods rather than on deductions from existing laws. He derived a set of "rules" for designing mills, including a critical examination of Smeaton's "maxims." He attempted to go even further and devised a set of rules for making inventions in any field. These amounted to applying scientific methods to technology. Included were the discovering of fundamental principles, the making of deductions from these principles, and the testing of the results by experiment.[10]

It is, of course, very difficult to determine which works were read by specific inventors; it is even harder to establish a correlation between particular inventions and previously published information. But it is easy to show that there was a vast increase in the volume of systematic technical literature available to American inventors in the course of the nineteenth century.[11] This was part of a world-wide movement that had its origins in the great encyclopedias of the eighteenth century. Oliver Evans, whose inventive career was pursued mostly before 1800, recalled that the chief impediment for the inventor at that time was the lack of reliable published information.[12] By the middle of the nineteenth century, through the efforts of men like Evans and Renwick, this barrier to invention had been largely removed.

Inventors might apply scientific methods; but, despite the work of

---

10. Oliver Evans, *The Young Mill-Wright and Miller's Guide* (Philadelphia, 1795), pp. 1–70, appendix pp. 1–2.

11. See Eugene S. Ferguson, *Bibliography of the History of Technology* (Cambridge, Mass., 1968); and Brooke Hindle, *Technology in Early America* (Chapel Hill, N.C., 1966).

12. Eugene S. Ferguson, ed., *Early Engineering Reminiscences of George Escol Sellers,* United States National Museum Bulletin 238 (Washington, D.C., 1965), p. 38.

a few like Evans, the inventor was ill-adapted to the task of building up the technological sciences. Scientists, on the other hand, had the necessary skills, and they played a vital role in stimulating the development of engineering sciences. But scientists lacked the lasting commitment and the intimate knowledge of technology and its needs that were required. The bulk of the effort to build technological sciences, therefore, fell on the engineering profession itself.

The engineering sciences, by 1900, constituted a complex system of knowledge, ranging from highly systematic sciences to collections of "how to do it" rules in engineering handbooks.[13] Some, like the strength of materials and hydraulics, built directly on science and were often classed as branches of physics. Others, such as the kinematics of mechanisms, evolved from engineering practice. In either case their development involved the adoption by engineers of the theoretical and experimental methods of science, along with many of the values and institutions associated with their use. By 1900, the point of origin made little difference; the engineering sciences constituted a unity. Those derived from practice took on the qualities of a science, in their systematic organization, their reliance on experiment, and in the development of mathematical theory. At the same time, sciences like the strength of materials gradually diverged from physics and assumed the characteristics of an autonomous technological science.

The separation of the engineering sciences from physics may be illustrated by the strength of materials and its sister disciplines, the theory of elasticity and the theory of structures. They were the first of the engineering sciences to be cultivated extensively in Europe or in America. The reasons for this were twofold. The intractable nature of materials constituted one of the most important barriers to

---

13. Hunter Rouse and Simon Ince, *History of Hydraulics* (Iowa City, Iowa, 1957); Eugene Ferguson, *Kinematics of Mechanism from the Time of Watt,* United States National Museum Bulletin 228 (Washington, D.C., 1962); Stephen P. Timoshenko, *History of the Strength of Materials* (New York and London, 1953); Isaac Todhunter, *A History of the Theory of Elasticity,* 3 vols. (New York, 1960). See also James H. Potter, ed., *Handbook of the Engineering Sciences,* 2 vols. (Princeton, N.J., 1967); and James K. Finch, "Engineering and Science: A Historical Review and Appraisal," *Technology and Culture,* II (Fall, 1961), 318–32.

the development of technology. The nineteenth century saw many new uses for materials like iron and steel; a scientific study of their properties would enable designers to avoid costly failures. But another reason for the early emphasis on this science was that it represented one of the most promising avenues for the application of science to technology. It could draw upon a sophisticated body of physics accumulated since the time of Newton. Thus, it attracted scientists and others inspired by the vision of a scientific technology.

European and American scientists played a key role in fostering the development of the science of the strength of materials. But once it was established, technologists dominated its further development, though scientists continued to make important contributions. Scientists such as Hooke, Euler, Young, and Coulomb did much to lay its foundations; it is worth remembering that the second of Galileo's "two new sciences" was the strength of materials. But once it had reached the stage of being technologically useful, engineers undertook its development. A critical institutional innovation was the development of engineering colleges in which technology would be pursued in the manner of science. The pioneer Ecole Polytechnique was widely imitated both in Europe and America. A group of polytechnicians, notably Louis Marie Navier and Barrie de Saint-Venant, reformulated and extended this science.[14]

As the strength of materials moved from the community of science to that of technology, it went through an important transformation. Its ties with physics were weakened, and it developed in ways uncharacteristic of the basic sciences. At the same time, its range of technological utility was gradually expanded. Scientists tended to explain their findings by reference to the most fundamental entities, such as atoms, ether, and forces. But these entities are not always directly observable. To be useful to a designer, however, a formulation must deal with measurable entities, particularly those of importance to the practical man. These need not be fundamental in the scientific sense. The scientists who had done so much to found a science of the strength of materials, notably Young, Coulomb,

---

14. Frederick B. Artz, *The Development of Technical Education in France, 1500–1850* (Cambridge, Mass., 1966), pp. 81–86, 151–66, 230–53; Timoshenko, *History of Strength of Materials,* pp. 67–80, 135–41, 229–42.

and Poisson, strove to found this study on the same ontological basis as classical mechanics, that is, they sought to explain their results in terms of molecules and the forces between them. Though not without interest, these efforts were not wholly successful. They were also needless complications from the technological point of view. A few of the engineers pioneering in this field, including Navier and Saint-Venant, continued this quest, but in the end the attempt was abandoned.[15] Instead, engineers were content with a simple macroscopic model—viewing a beam, for example, as a bundle of fibers.

In America, as in Europe, the foundation of the science of strength of materials owed much to scientists. Led by Alexander D. Bache, the Franklin Institute in 1830 undertook an investigation of the causes of steam-boiler explosions for the federal government. This study was itself one of the first significant attempts to use scientific methods to investigate technological problems in America.[16] One aspect of this multifaceted effort was a systematic study by Walter R. Johnson of the strength of the metals used in boiler construction. This involved building the first testing machine in America and conducting a well-conceived and highly fruitful series of experimental tests.[17] Scientists also fostered the use of mathematical theory for the study of materials. William Barton Rogers, though primarily a geologist, was well grounded in physics and mathematics. His *An Elementary Treatise on the Strength of Materials,* published in 1838, was the first American book in this field.[18]

Perhaps the greatest contribution of scientists like Rogers was their fostering of institutions to encourage the marriage of science

---

15. Timoshenko, *History of Strength of Materials,* pp. 104–7, 231–32.

16. Bruce Sinclair, *Early Research at the Franklin Institute: The Investigation into the Causes of Steam Boiler Explosions, 1830–1837* (Philadelphia, Pa., 1966); John G. Burke "Bursting Boilers and the Federal Power," *Technology and Culture,* VII (Winter, 1966), 1–23.

17. "Report of the Committee of the Franklin Institute . . . on the Explosion of Steam Boilers . . . Part II . . . ," *Journal of the Franklin Institute,* XIX (February, 1837), 73–109; (March, 1837), 156–93; (April, 1837), 241–77; (May, 1837), 28–31; (June, 1837), 409–51; XX (July, 1837), 1–31; (August, 1837), 72–113. See also George E. Pettengill, "Walter Rogers Johnson," *Journal of the Franklin Institute,* CCL (August, 1950), 93–113.

18. William Barton Rogers, *An Elementary Treatise on the Strength of Materials* (Charlottesville, Va., 1838).

and technology. In the forefront, Rogers had apparently become concerned with technology when he lectured at the Maryland Institute, a Baltimore mechanics' institute, in 1827. His treatise on the strength of materials was produced as part of an effort to found a school of engineering at the University of Virginia. While this venture did not succeed, Rogers did not give up. In 1846, he drew up a plan for a "polytechnic school" for Boston, a dream finally realized in 1861 with the chartering of the Massachusetts Institute of Technology.[19] Rogers was, of course, not alone in his vision of a scientific technology. Headed for the most part by chemists and geologists, the scientific schools attached to Harvard, Yale, and other colleges instituted engineering programs. Benjamin F. Green reorganized Rensselaer into a polytechnic school after 1847. Rensselaer was one of the first to concentrate almost exclusively on engineering, and the first to go beyond one-man departments in this area, a vital step in encouraging the specialization required for the development of the engineering sciences.[20]

While scientists like Rogers, Bache, Renwick, and Green did much to found the scientific study of materials in America, its systematic development lay principally with the engineers themselves. An important role in this was played by West Point, the first American engineering school. It was reorganized after 1818 by Sylvanus Thayer on the model of the great French engineering schools. One graduate, Dennis Hart Mahan, was sent to France to complete his engineering education; on his return, he taught civil and military engineering to cadets from 1832 to 1871. In 1837, Mahan produced the first American textbook based on French engineering practice, *An Elementary Course of Civil Engineering*. Over 15,000 copies of this work were sold; and the book had an important impact on the teaching of engineering in America.[21] It

19. Emma Rogers, *Life and Letters of William Barton Rogers*, 2 vols. (Boston and New York, 1896), I, 40–54, 259–62, 420–27.

20. Samuel Rezneck, *Education for a Technological Society* (Troy, N.Y., 1968), pp. 78–110; Palmer C. Ricketts, *History of the Rensselaer Polytechnic Institute* (New York, 1895), pp. 69–112.

21. George W. Cullum, "Dennis H. Mahan," *Biographical Register of the Officers and Graduates of the U.S. Military Academy at West Point*, 7 vols. (Boston and New York, 1891), I, 319–25.

included a brief survey of the strength of materials. Although Mahan limited himself to elementary mathematics, his treatment of this subject was distinctly professional, in striking contrast to the purely qualitative discussion of the strength of materials in Bigelow's *Elements of Technology*. It is perhaps significant that while Bigelow sought explanations for the properties of materials in molecules and forces between them, Mahan made no reference to these fundamental entities of physics. Mahan was well read in the European literature, and he particularly recommended the works of Navier to his students. A few apparently followed his advice. West Point engineers did much to establish a tradition of the scientific study of engineering in America.[22]

Mahan's work and the technical works which followed it provided a basis for introducing European methods into ordinary engineering practice in America, but creative contributions, the founding of a science, required money for laboratories and equipment as well as men trained to use them. The federal government played an important role in supporting experimental investigations in the second quarter of the nineteenth century. Federal funds had made possible the Franklin Institute's studies of boiler explosions. The same testing machine was used for another pioneering investigation, the study of the causes of the disastrous explosion of a cannon on the U.S.S. *Princeton*.[23] But the institute lacked the funds for developing a sustained program of research. Army officers, particularly in the Ordnance Department, to some extent filled in the gap. A series of experiments on the strength of cannons by Major William Wade and Captain Thomas Jackson Rodman were among the first American contributions to attract European attention. Wade's testing machine was apparently the second to be built in America.[24] Other

---

22. Dennis H. Mahan, *An Elementary Course of Civil Engineering* (New York, 1837), pp. vii, 44–53, 86–104; Bigelow, *Elements of Technology*, pp. 43–53.

23. Lee M. Pearson, "The Princeton and the Peacemaker: A Study in Nineteenth-Century Naval Research and Development Procedures," *Technology and Culture*, VII (Spring, 1966), 163–83.

24. U.S. Ordnance Department, *Reports of Experiments on the Strength and Other Properties of Metals for Cannon, with a Description of the Machines for Testing Metals, . . .* (Philadelphia, Pa., 1856). On Wade's testing machine see Chester H. Gibbons, *Materials Testing Machines* (Pittsburgh, Pa., 1935), pp. 27–28. For a critique of the work of Wade and Rodman, see Todhunter, *History*

experimental investigations were carried on by John Dahlgren and Benjamin F. Isherwood of the navy.[25] But the government was unwilling to make a long-range commitment for research not directed to an immediate mission. In 1872, the American Society of Civil Engineers requested that the government undertake tests of the properties of American iron and steel. Congress authorized a study and created a board of seven engineers to supervise the work. However, the president placed the control of the program with the Ordnance Department, which differed with the civilian engineers. In 1878, Congress turned the testing machine over to the army and dissolved the board. The army, on the grounds that they lacked the necessary funds, refused to cooperate with the civilian engineers.[26]

Large business ventures were also in a position to undertake scientific studies. The proprietors of Lowell supported James Francis' hydraulic experiments, but for his studies of the strength of cast iron he had to rely on European data.[27] The building of the Eads bridge necessitated the adoption of systematic testing of materials. This practice gradually spread through the steel industry, but these tests were usually geared to the needs of particular projects.[28] Thus, while business and government did much to encourage the adoption of experimental methods in technology, they were unwilling to carry out basic research on a sustained basis.

Engineering needed not just short-term studies directed to specific

of the Theory of Elasticity, II, pt. 1, 688–96. For Rodman's later work, see U.S. Ordnance Department, *Reports of Experiments on the Properties of Metals for Cannon, and the Qualities of Cannon Powder . . . by Captain T. J. Rodman* (Boston, 1861).

25. Edward William Sloan, *Benjamin Franklin Isherwood, Naval Engineer* (Annapolis, Md., 1965); Clarence S. Peterson, *Admiral John A. Dahlgren, Father of U.S. Naval Ordnance* (New York, 1945). Both Isherwood and Dahlgren were active in applying experimental methods to derive design principles for engineering. Isherwood was concerned with the design of marine steam engines, the design of screw propellers, and other subjects. Dahlgren, along with Rodman, used experiment to design the bottle-shaped cannon used in the Civil War.

26. Charles W. Hunt, *Historical Sketch of the American Society of Civil Engineers* (New York, 1897), pp. 82–83; William F. Durand, *Robert Henry Thurston* (New York, 1929), pp. 79–81.

27. James B. Francis, *Lowell Hydraulic Experiments* (Boston, 1855), p. xi; idem, *On the Strength of Cast-Iron Pillars* (New York, 1865), pp. 1–17. Francis derived a classic set of design principles for turbines in the former work (pp. 44–52).

28. Carl W. Condit, *American Building Art: The Nineteenth Century* (New York, 1960), pp. 9, 139–40; Gibbons, *Materials Testing Machines*, pp. 31, 34–40.

problems, but a broad, continuous program of basic research in laboratories specifically dedicated to developing the engineering sciences. Robert Thurston, one of the founding fathers of mechanical engineering in America, was perhaps the foremost champion of basic research in the engineering sciences. He wanted laboratories established in connection with engineering schools. The rise of research-oriented universities and technical institutes after the Civil War gave him his opportunity. He founded two of the earliest and best-known engineering laboratories in the United States at Stevens Institute of Technology and Cornell University. Thurston devised two new testing machines and made important discoveries of the properties of materials. With his three-volume work, *The Materials of Engineering*, the experimental study of the strength of materials reached maturity in America.[29]

Though the experimental approach to technology was readily adopted in America, theory tended to lag behind. American technologists generally lacked the advanced mathematical training needed to make contributions to a sophisticated field like the theory of elasticity. American engineers also tended to pride themselves on their practicality; they regarded mathematical theory as of little real value. The theoretical approach had to prove its utility to be adopted. The difficulty lay with the limitations of existing theory. Though the strength of materials had developed into a science by the 1830s, the range of application of its theory was limited. Very elegant solutions for a limited number of problems were available; but most problems were not solvable.[30] Many problems were indeterminate; they could not be solved because the number of unknowns was greater than the number of equations. Unfortunately, the indeterminate cases included some of the ones most frequently met in American engineering practice: the continuous beam and the truss bridge.[31]

---

29. Durand, *Thurston*, pp. 65–73, 236–40; Robert H. Thurston, "On the Necessity of a Mechanical Laboratory," *Journal of the Franklin Institute*, LXX (December, 1875), 409–18; idem, *The Materials of Engineering*, 3 vols. (New York, 1883).
30. Timoshenko, *History of Strength of Materials*, p. 231.
31. Condit, *American Building Art*, pp. 6–9.

From the 1830s to the 1870s, there was a major effort in Europe and in America to extend the range of applicability of the engineering sciences. This effort met with remarkable success; by 1880 it was possible to attack a wide range of problems by mathematical theory. In America much of the effort went into the analysis of truss bridges. Squire Whipple's *An Elementary and Practical Treatise on Bridge Building*, which first appeared in 1847, was a homespun product developed apparently in complete innocence of European work. Whipple employed no mathematics other than elementary geometry and algebra; he did not use calculus or even trigonometry. While he expressed his results algebraically, the argument was basically geometrical, giving his work a quaint, seventeenth-century flavor at times. Nevertheless, Whipple's work was a remarkable achievement. He derived mathematical and graphical methods by which he was able to analyze correctly truss bridges which were indeterminate by the usual methods.[32]

A second American effort to establish a mathematical theory for bridges was that of Herman Haupt, whose *General Theory of Bridge Construction* appeared in 1851. A West Point graduate, Haupt had some familiarity with European theory. The British scientist Thomas Young, upon whose work Haupt relied heavily, assumed, like many scientists, that stresses were ultimately reducible to forces between particles. On this assumption, Haupt sought to resolve the forces operating on a beam to a single resultant force acting at the center of an equivalent geometric figure. Unfortunately, stresses are not forces, and they cannot be combined in this manner. Though Haupt's assumptions were open to question, his approximations were doubtless a vast improvement over the rule-of-thumb methods still in general use. A correct theory of stresses, developed at about the same time by European engineers, did much to further the separation between engineering sciences and physics. It was no longer helpful to attempt to base this engineering science on atoms and forces, the fundamental assumptions of physics.[33]

---

32. Squire Whipple, *An Elementary and Practical Treatise on Bridge Building*, 4th ed. (New York, 1883). See also *idem*, "On Truss Bridge Building," *Transactions of the American Society of Civil Engineers*, I (1868–71), 239–44.

33. Herman Haupt, *General Theory of Bridge Construction* (New York, 1851). Haupt's assumption was that "the weight of any body may be supposed

Whipple's and Haupt's use of graphical methods to extend the range of engineering science was prophetic of one of the principal advances in the strength of materials studies. In 1866, the Swiss engineer Karl Culmann developed an important graphical method in which stresses were represented by segments of circles. Culmann's use of circle diagrams was extended by the German engineer Otto Mohr and others and resulted in a great increase in the range of usefulness of the strength of materials.[34] The development of graduate-level work at American universities after the Civil War produced engineers who had the training to develop and apply methods of mathematical theory to materials. Henry Turner Eddy, a graduate of Yale's Sheffield Scientific School, who received his Ph.D. from Cornell, was among the first of this new generation of scientific technologists. In 1878, he published an extension of the new graphical methods in his *Researches in Graphical Statics*. It was one of the first American engineering books to be translated into German; Florian Cajori, the historian of American mathematics, called it "the first original work on this subject by an American writer."[35]

The expanded range of application of the engineering sciences was accompanied by a tendency away from analytic solutions, a reliance on approximations, and, to some extent, a lessening of

---

concentrated at its center of gravity; and, in general, any number of parallel forces may be replaced by a single force called the resultant. In the present case . . . the sum of all the forces upon the fibres . . . will be the same, as if a single force equal to its area was applied in the direction of a line passing through its center of gravity" (p. 20). Each normal stress is always accompanied by two components of shearing stress acting at right angles. These cannot be combined by a parallelogram of forces to give a single resultant. In modern terms, forces behave like vectors, but stresses behave like tensors. See also Thomas Young, *A Course of Lectures on Natural Philosophy and the Mechanical Arts* (London, 1807), pp. 135–52.

34. Timoshenko, *History of the Strength of Materials,* pp. 190–97, 283–88. See also Hans Straub, *A History of Civil Engineering* (Cambridge, Mass., 1964), pp. 197–202. Not all of the changes were in the direction of lessening rigor; Saint-Venant was opposed, and his development of the "semi-inverse" method extended rigorous analytic solutions.

35. Florian Cajori, *The Teaching and History of Mathematics in the United States* (Washington, D.C., 1890), p. 177; Henry Turner Eddy, *Researches in Graphical Statics* (New York, 1878); and *idem, Neue Constructionen aus der graphischen Statik* (Leipzig, 1880). See also Arthur E. Haynes, "Henry Turner Eddy," *The Minnesota Engineer,* XX (March, 1912), 104–7; and *Dictionary of American Biography,* s.v. "Henry Turner Eddy."

mathematical rigor. A given problem in the strength of materials might be solved rigorously by the theory of elasticity, or it might be treated by less rigorous graphical methods. American engineers, beginning with Rodman, pioneered even less rigorous empirical methods, using strain gauges and models. The selection of technique depended on economic as well as technical factors, since rigorous treatment, when possible, often involved more time and effort. The development of hierarchies of methods of variable rigor, along with the importance of economic factors in determining their use, served to distinguish the engineering sciences from physics, where only the most rigorous methods were normally admitted.[36]

By 1900, the American technological community was well on the way to becoming a mirror-image twin of the scientific community. The rise of engineering sciences had played a vital role. They gave technology equivalents to the theoretical and experimental departments of physical science. They were fostered by engineering colleges, which, by 1900, had virtually displaced apprenticeship as a means of training engineers. Scientifically inclined engineers like Thurston played an important role in the founding of professional engineering societies after the Civil War and, even more important, in producing worthwhile technical literature for engineering journals. But despite the structural similarities between science and technology, the two were further apart in some respects. In many important areas, engineering and physics had ceased to speak the same language.

In the case of mirror-image twins, there is a subtle but irreconcilable difference which is expressed as a change in parity. Between the communities of science and technology, there was a switch in values analogous to a change in parity. One way of putting the matter would be to note that, while the two communities shared many of the same values, they reversed their rank order. In the physical sciences, the highest prestige went to the most abstract and general,

---

36. Eddy's and Mohr's methods rested on rigorous mathematical proofs. But graphical methods usually involved simplifying assumptions about the distribution of stresses. For Rodman's pressure gauge, see U.S. Ordnance Department, *Reports of Experiments . . . by Captain T. J. Rodman*, pp. 299–300.

that is, to the mathematical theorists from Newton to Einstein. Instrumentation and applications generally ranked lowest. In the technological community, the successful designer or builder ranked highest, the "mere" theorist the lowest. These differences are inherent in the ends pursued by the two communities: scientists seek to know; technologists, to do. These values influence not only the status of occupational specialists, but the nature of the work done and the language in which that work is expressed.

An indication of the gap between science and technology is provided by comparing two discoveries, one by the American physicist Henry Rowland and the other by Francis Hopkinson, a British electrical engineer. Rowland, starting from an idea of Faraday's, published a paper on magnetic permeability in 1873. James Clerk Maxwell, to whom Rowland sent the paper, recognized its importance, and arranged to have it published in *Philosophical Magazine*. In 1879, Hopkinson published the results of his investigation of the efficiency of electric dynamos. By graphing his results, he discovered the "characteristic curve" of the direct-current dynamo, a vital key to rational design. Hopkinson was able to show, for example, how the Edison dynamo could be radically improved by simply changing the dimensions of some of its parts. It was not discovered until several years later that, in a certain sense, Rowland and Hopkinson had made the same discovery.[37]

There was an irony in the fact that Rowland had "discovered" a key to the design of electric dynamos without realizing it, for his only earned degree was in engineering. While he had transferred his primary loyalty to physics, his laboratory at Johns Hopkins was an important center for the training of electrical engineers. Rowland missed the significance of his discovery because he was looking for a law of nature, not a design principle. In the case of Rowland and Hopkinson, each expressed his work in the terms appropriate to his quest: Rowland discovered a relation between the entities of electromagnetic theory; Hopkinson found one between basic engi-

---

37. James E. Brittain, "B. A. Behrend and the Beginnings of Electrical Engineering, 1870–1920" (Ph.D. diss., Case Western Reserve University, 1969), pp. 6–18.

neering parameters, such as the input and output of a dynamo. The method of approach, the argument, and the form of presentation differed according to the purpose and the audience for which the results were intended. The two might be considered equivalent because the engineering variables of Hopkinson could be expressed as functions of the electromagnetic entities employed by Rowland.[38]

Perhaps no scientist has had a greater impact on technology than James Clerk Maxwell, but his influence was indirect, since few engineers could understand him. It required a creative effort almost equal to Maxwell's own by the British engineer Oliver Heaviside to translate his electromagnetic equations into a form usable by engineers.[39] Yet Maxwell was one of those scientists who consciously attempted to contribute to technology. Thus, he developed an important method for solving indeterminate problems in the theory of structures. But this work, too, had to be translated for technologists. A British engineer, after quoting Maxwell's conclusions, commented that "few engineers would, however, suspect that the two paragraphs quoted put at their disposal a remarkably simple and accurate method of calculating the stresses in a framework."[40]

The cases of Rowland and Maxwell suggest how the interchange between science and technology may take place. The passage of information from one community to the other often involves extensive reformulation and acts of creative insight. This requires men who are in some sense members of both communities. These intermediaries might be called "engineer-scientists" or "scientist-engineers," depending on whether their primary identification is with engineering or with science. Such men play a very important role as channels of communication between the communities of science and technology. It is significant that Joseph Henry, Alexander D. Bache, Henry Rowland, and J. Willard Gibbs were all trained as engineers. Administrators of scientific agencies of government and those engaged in teaching science to engineers could be more effective if

---

38. *Ibid.*, p. 41.

39. James E. Brittain, "Heaviside and the Telephone: A Case Study of the Interaction of Science and Technology in Nineteenth-Century Telephony" (Master's essay, Case Western Reserve University, 1968), pp. 1–3, 7–14. See also Oliver Heaviside, *Electrical Papers*, 2 vols. (Boston, 1925).

40. Quoted in Timoshenko, *History of Strength of Materials*, p. 203.

they were capable of understanding and reconciling the competing demands of science and technology.

It is worth noting, however, that the relationship between science and technology is a symmetrical one. That is, information can be transferred in either direction. The flow of technology into science in the form of instrumentation has long been recognized; but the traditional model does not provide for the possibility that technological theory might influence science. The rise of engineering sciences such as the theory of elasticity and hydrodynamics, however, did have an influence on science. The theory of elasticity provided a means of constructing models of the ether, a favorite occupation of Lord Kelvin and others. The maturing of hydrodynamics was one cause of the proliferation of vortex theories of matter in the second half of the nineteenth century. Thermodynamics was the product of a somewhat more complex interaction. This science began as a design principle of the French engineer Carnot. It was not a law of nature but a statement of the limits of the efficiency of the steam engine. Its development relied on a simple macroscopic model, Carnot's ideal heat engine; it did not rely on molecular hypotheses. It was discovered in the engineering literature by scientist-engineers like Kelvin, Rankine, and Helmholtz, and was translated by them into the language of science. As thermodynamics was absorbed by physics, Carnot's ideal heat engine was replaced by the molecular model of statistical mechanics.

Because of the status differentials, one would expect engineers with the appropriate training to attempt some work that was directed at the scientific community. Theory ranks high in science, but low in engineering. Many examples of American engineers contributing to basic science could be cited. DeVolson Wood used elasticity considerations in an attempt to determine the density, pressure, and specific heat of the ether. Though one of the weakest of his works, Wood apparently took inordinate pride in it and published an expanded version as a book. Henry Turner Eddy produced some interesting papers in which he concluded from kinetic considerations that the atom must have some form of internal motion and he postulated the existence of a subatomic particle.[41] The only earned degrees of J. Willard Gibbs and Henry Rowland were in engineering. Considering the low status of academic theorists in engineering, not to mention the low status of engineers as a group, their identification with physics is not surprising.

The most important influence of technology on scientific ideas, however, was more indirect. Engineering sciences did not postulate unobservables. Their example was, therefore, a challenge to physics. They contributed to the critical reexamination of the foundations of physics which took place in the late nineteenth century. But the engineers themselves contributed little to this movement; it was carried forward by physicists under the banners of positivism and energeticism. The influence of technology on science, like that of science on technology, was an indirect, second-order effect.

The coupling of science and technology in the nineteenth century had at least two important social consequences in the twentieth century. It accelerated the pace of technological change and, consequently, of social dislocation. It also encouraged engineers to adopt a self-image based on science, which served to discourage them from assisting society in meeting the problems they had done so much to create. The scientific self-image caused engineers to portray themselves as logical thinkers, free of all bias and emotion, and it promoted an "above-the-battle" neutrality on the part of the profession. Though engineers gave lip service to the idea of social responsibility, their definition of this responsibility served to prevent effective action. When faced with an actual social problem, engineers sought objective or "scientific" solutions. In practice, they set the discovery of methods of social engineering as a precondition to social action, thus substituting an impossible task for a difficult one. (The delusive quest for social engineering led more than one engineer down the blind alley of technocracy.)

The reversal of "parity" between science and technology further

---

41. David R. Topper, "The Development of the Kinetic Theory of Gases in America: An Analysis of the Ideas of Three Key American Figures Prior to Gibbs" (Master's essay, Case Western Reserve University, 1968), pp. 35–62. DeVolson Wood, *The Luminiferous Aether* (New York, 1886).

reduced the engineers' ability to respond effectively to social problems. The scientific community was better able to act on social issues because those with the greatest prestige were in universities, where they were relatively free from pressures from corporations and government. The engineers who enjoyed a corresponding independence lacked sufficient prestige to lead their profession. Prestige and power in engineering went to the "doers," not to the "theorists." This had the practical effect of giving the control of the engineering profession to men who were linked by ties of self-interest to those who were using, and in some cases misusing, technology. This conflict in interest between the leaders of the profession and the rank-and-file engineers did much to frustrate the legitimate professional aspirations of American engineers.[42]

42. Cf. Edwin T. Layton, Jr., *The Revolt of the Engineers: Social Responsibility and the American Engineering Profession* (Cleveland, Ohio, 1971).

Science and Industry

MODERN INDUSTRIAL RESEARCH

In 1966, 71 per cent of the nearly one and a half million scientists and engineers living in the United States were employed in private industry.[1] So intimately are science and industry connected in the popular mind that we are beginning to speak of a second Industrial Revolution. Industrial research laboratories, such as those maintained by General Electric and by American Telephone and Telegraph, are so large and so productive that progress itself has become an object of the hucksters' art. It is not surprising, therefore, that historians have sought to discover from whence has come this dramatic efflorescence of applied science.

The earliest, and in some ways still the best, such effort was made by Howard R. Bartlett in the 1930s. "The Development of Industrial Research in the United States," by Bartlett, chronicles the progressive removal of impediments to and the marshaling of support for the conduct of organized research within industry. "Not until the [1890s]," he wrote, "had the developments in science, education, and industry reached the point at which the organized application of science to industry by trained men seemed to industrialists to be the key to greater progress and profit."[2] He felt that the most signifi-

---

1. National Science Foundation, *Employment of Scientists and Engineers in the United States, 1950–66* (Washington, D.C., 1968), p. 8.

2. Howard R. Bartlett, "The Development of Industrial Research in the United States," *Research—A National Resource*. II: *Industrial Research*, Report of the National Research Council to the National Resources Planning Board, December, 1940 (Washington, D.C., 1941), p. 19.

cant developments in industry were the growing scarcity of natural resources that could be easily exploited and the consequent importance of efficient, economical methods of production to sustain profits.

Developments in education centered around the growth of graduate schools such as Johns Hopkins, which emphasized academic research, and technical schools such as the Massachusetts Institute of Technology, which turned out increasing numbers of scientifically trained students. The most important development within science itself was its ever increasing accumulation and control of data: "The reservoir of scientific knowledge," wrote Bartlett, "was filling."[3] The culmination of all these progressive developments was, of course, the establishment of industrial research laboratories early in the twentieth century.[4]

More recently Kendall Birr has also turned to this subject. In his essay "Science in American Industry," Birr describes the origins of the industrial research laboratory.[5] Most industries in the nineteenth century, he maintains, were based on "relatively primitive technologies."[6] Gradually, with "the proliferation of scientific knowledge and the establishment of basic systems of interpreting the data," science began to provide specific solutions to a limited number of industrial problems. More important, a few industries were science-based from their beginnings. The electrical industry and certain chemical industries were examples. Birr writes:

By the end of the century, many American industrial firms were using scientifically trained men; their inventions were being exploited by industry, scientists were widely employed to analyze and control existing processes, and they were frequently called on as consultants. The time was ripe for the introduction of the modern industrial research laboratory.[7]

---

3. *Ibid.*, p. 20.

4. A. D. Little and H. E. Howe, "The Organization and Conduct of an Industrial Laboratory," *Transactions of the American Society of Mechanical Engineers*, XLI (1919), 69.

5. Kendall A. Birr, "Science in American Industry," *Science and Society in the United States*, ed. David D. Van Tassell and Michael G. Hall (Homewood, Ill., 1966), pp. 35–80.

6. *Ibid.*, p. 37.

7. *Ibid.*, p. 68.

This concern with the history of the industrial research laboratory was reinforced by Birr's excellent history of the General Electric laboratory. The study looks back at the nineteenth century to find the roots of modern laboratories in German universities and in a few industrial concerns. Thomas Edison's Menlo Park establishment, for example, had twenty men at work as early as 1878.[8] John Beer and W. David Lewis, in their study of the professionalization of science, have discussed some of the problems of early industrial laboratories, and reports of these problems have punctuated the writings of applied scientists in the early twentieth century.[9]

In the few studies of the history of industrial research, certain aspects of the subject have received more attention than have others. In part, no doubt, this disparity reflects the value system of the scientific profession itself: pure research is better than applied research, and any kind of research is better than mere analysis or application. Discussions of the subject of "science and industry" easily become bogged down in such distinctions as that between "science" and "scientist," or between "true science" and research which has since proved to be in error. Too often we have tended to dismiss men who were "not really scientists," or who tried to apply "wrong" scientific principles, or who, at any rate, were not successful in producing innovations. If there was no "real" science in nineteenth-century America, by definition there could not possibly be any significant relationship between science and industry. The danger of such an approach, of course, is that it leads us to overlook a good deal of interesting, and possibly significant, activity.

Quite as dangerous as a too narrow definition of science is an unrealistic conception of the role science plays today in technological innovation. Scientists' arguments that pure research brings industrial rewards tempt us to see science as the only, or at least the major, source of innovation. Recent studies, however, cast some doubt on the assumption that the discovery of new natural laws in scientists' laboratories ought to lead rapidly to the establishment of new in-

---

8. Kendall Birr, *Pioneering in Industrial Research: The Story of the General Electric Research Laboratory* (Washington, D.C., 1957), pp. 7–8.

9. John J. Beer and W. David Lewis, "Aspects of the Professionalization of Science," *Daedalus*, XCII (Fall, 1963), 764–84.

dustries, the marketing of new products, or the improvement of established processes. Not only did this fail to happen in the nineteenth century, but it apparently does not happen even today when the relationship between science and industry is presumably elaborate and mature.

SOURCES OF INNOVATION

During the 1960s three major studies, one sponsored by the Department of Defense and two by the National Science Foundation, came to different conclusions concerning the importance of basic research, but otherwise reveal strikingly similar patterns of technological innovation.[10] The picture that emerges from these studies stands in dramatic contrast to the "better things for better living through chemistry" image of the industrial research laboratory which we have inherited from the 1920s. There seems to be general agreement that technological innovations are most often the result of accumulated small technical changes and are usually dictated by an intimate knowledge of market or production needs. When new scientific information is added, it comes most often from the experience and education of the men involved and only rarely from the current scholarly literature. When a piece of basic research is involved, it is typically two decades or more old before it shows up in commercial application. The chances for successful innovation would appear to be at their best when a number of men trained and experienced in the sciences are intimately involved in making and marketing commercial products or services.

What is true of our own century may not, of course, have been true of the nineteenth century. The rise of the large industrial research laboratory has made important changes in some areas, and

---

10. Chalmers W. Sherwin and Raymond S. Isenson, "Project Hindsight," *Science,* CLVI (June 23, 1967), 1575; D. S. Greenberg, " 'Hindsight': DOD Study Examines Return on Investment in Research," *Science,* CLIV (November 18, 1966), 872–73; *Technology in Retrospect and Critical Events in Science,* prepared for the National Science Foundation by the Illinois Institute of Technology (Washington, D.C., 1968); and Sumner Myers and Donald G. Marquis, *Successful Industrial Innovations: A Study of Factors Underlying Innovation in Selected Firms* (Washington, D.C., 1969).

the presence of over a million scientists and engineers, three-quarters of them busy in private industry, represents a striking quantitative change in this century. It is still possible, however, and in fact it seems plausible, that the general conditions which tend to maximize innovation today are not structurally different from what they were a century ago. Indeed, John Jewkes and his associates have concluded that, in terms of inventions,

> There are many similarities between the present and the past century in the type of men who invent and the conditions under which they do so. Many of the twentieth-century stories could be transplanted to the nineteenth without appearing incongruous to the time or the circumstances; far too many, indeed, to render tenable the idea of a sharp and complete break between the periods.[11]

Fresh insight into that time and those circumstances has been provided by the sociologist Joseph Ben-David. In seeking the nature and cause of the apparent disparity between the support and use of science in the United States, as compared with Western Europe, he was led to the conclusion that the roots of the disparity lay in the nineteenth century. "Success in exploiting science for practical purposes," writes Ben-David, "does not . . . result from the guidance of fundamental research by practical considerations, but from constant entrepreneurial activity aimed at bringing to the attention of potential users whatever may be relevant for them in science, and *vice versa.*" [12] In the nineteenth century, branches of industry were much more numerous than the fields of science, so it was usually easier for an industrialist to identify the potentially useful science than for the scientist to see which industry might most profitably use his research.

Nevertheless, the flow was both ways, and if not all scientists or manufacturers were anxious to work together, American society was full of institutional entrepreneurs, from textile magnates to

---

11. John Jewkes, David Sawers, and Richard Stillerman, *The Sources of Invention* (London, 1961), pp. 89–90.

12. Joseph Ben-David, *Fundamental Research and the Universities: Some Comments on International Differences* (Paris, 1968), p. 56.

college presidents, who were anxious to effect such a partnership. Ben-David reports that the institutions of higher education were of critical importance. Already in nineteenth-century America, captains of erudition, more distinguished for their entrepreneurial skill than their scholarship, were founding the knowledge industry which has so distinguished our own time. Increasingly, colleges contributed both new technical information and a host of students trained in application of this information to social needs.[13]

PRINCIPLES FOR THE ARTS TO APPLY

The blocks to the union of science and the useful arts were real, of course, as Bartlett, Birr, and others have demonstrated. There was, for example, a strong feeling on the part of some scientists that thought was better than action, and that their place was in the laboratory, not the market. Addressing the American Institute in 1851, the controversial Dr. Charles T. Jackson asserted that

> No true man of science will ever disgrace himself by asking for a patent; and if he should, he might not know what to do with it any more than the man did who drew an elephant at a raffle. He cannot and will not leave his scientific pursuits to turn showman, mechanic, or merchant; and it is better for him and for the world that he should continue his favorite pursuits and bring out more from the unexplored depths of human ingenuity and skill.[14]

Nevertheless, the conventional wisdom in the early 1800s was that science and industry were related; as Alonzo Potter put it in 1841, "On the one hand, science has furnished principles for the arts to apply: on the other hand, the arts have proposed problems for science to resolve; and this mutual aid and dependence have been the means of carrying both forward at a rate continually accelerated." Even though "the useful arts may precede science, at first, they will subsequently follow and be guided by its light."[15] Joseph Henry

---

13. *Ibid.*, pp. 38, 45–46.
14. *Scientific American*, VII (November 1, 1851), 51.
15. Alonzo Potter, *The Principles of Science Applied to the Domestic and Mechanic Arts, and to Manufactures and Agriculture* (Boston, 1841), 266–67, 12.

expressed a common sentiment in 1826 in the context of his inaugural lecture as professor of mathematics and natural philosophy at the Albany Academy. "One great object of science," he declared, "is to ameliorate our present condition, by adding to those advantages we naturally possess." Furthermore, he went on, "In nothing do mathematical and philosophical principles appear more decidedly useful than in their application to the mechanic arts. To these they present in a condensed form the united experience of many ages; by a combination of theoretical knowledge with practical skill, machines have been constructed no less useful in their productions than astonishing in their operations." [16]

Rarely was this dogma explicitly called into question. One occasion was an address by Daniel Treadwell, practical inventor and Rumford Professor and Lecturer on the Application of Science to the Useful Arts at Harvard College. Speaking before the engineers of the American Academy of Arts and Sciences in 1852, Treadwell took particular aim at Francis Bacon and the idea that he had established, "in the inductive system of philosophy, the true art of invention." He insisted that "the improvement of the useful arts has never at any time been the vocation of the philosophers," and singled out the efforts of Sir Humphrey Davy as a particularly clear example of "the imbecility of science in improving the arts." If the inductive method meant the pure Baconian regimen, Treadwell denied that any scientist had ever really used it to good purpose even in the field of science. If it meant merely arriving at general principles and laws by reasoning from particular instances, he claimed that ancient men as well as modern ones, in and out of schools, had often operated in this way. "Before Bacon taught him otherwise," scoffed Treadwell, "it is said that man rested entirely, for what he knew, upon dialectics and logic. As though there was no knowledge out of the schools! Were the Pyramids reared and edified by logic?" [17] What Treadwell denied was not the efficacy of science, but the philosophers' exclusive claim to scientific practice.

---

16. *Albany* (N.Y.) *Argus and City Gazette*, September 18, 1826.

17. Daniel Treadwell, *The Relations of Science to the Useful Arts: A Lecture Delivered to the American Academy of Arts and Sciences, November, 1852* (Cambridge, Mass., 1855), pp. 10–11, 15–16, 24–25.

In fact, Treadwell had isolated an important source of confusion. A double standard was being applied to the problem of science and the technical arts. When a scientist conducted experiments in his laboratory, and then deduced general principles from that experience, he was practicing science in the best Baconian tradition. When the mechanic or engineer did the same thing, his activity was termed cut-and-try, and his conclusions were the result of mere accumulated experience. Even illuminated by prejudice, however, the distinction was not always clear. What was one to make of Henry's statement that mathematics and philosophy "present in a condensed form the united experience of many ages"? Many an American mechanic of the nineteenth century undoubtedly considered himself something of a scientist when he conducted experiments on machinery or industrial processes and then tried to account for the results in terms of what were usually called "principles." He would no doubt have been confused had he been told that the scientist was truly scientific only when he was being inductive, but that for the mechanic or manufacturer to be scientific he must be deductive, that is, make particular application of already established principles.

Oliver Evans, in many ways the greatest American inventor of his time, took principles very seriously; he applied them when they were available and sought them when they had not yet been established. In an attempt to find utility in current science, he noted in the postscript to his handbook, *The Abortion of the Young Steam Engineer's Guide,* that "chemists, to prevent the confusion of ideas, have invented a new name for the substance of heat, which they call *Caloric,* to distinguish it from the effects which it produces." He pointed out that "if there be any substance through which it cannot pass, the discovery thereof may prove eminently useful in the arts, especially in the construction of furnaces for steam engines." [18]

In describing his own work, Evans claimed that "it has been by the most intense study that I have made discoveries. After having a faint glimpse of the principle, it was with many toilsome and tedious steps that I attained a clear and distinct view. I received

---

18. Oliver Evans, *The Abortion of the Young Steam Engineer's Guide* (Philadelphia, Pa., 1805), p. 134.

great assistance from the result of experiments made by others, which are to be found in scientific works." [19] He had, for example, followed the work of Dalton but concluded in this case that the English chemist's work had been in error. Although Evans himself had caught the error, he claimed that the steamboat inventor John Stevens, of Hoboken, had accepted Dalton's work and been misled by it.[20] So important did Evans consider this sort of work that he advocated a program of federal subvention of scientific research:

> If government would, at the expense of the community, employ ingenious persons, in every art and science, to make with care every experiment that might possibly lead to the extension of our knowledge of principles, carefully recording the experiments and results so that they might be fully relied on, and leaving readers to draw their own inferences, the money would be well expended; for it would tend greatly to aid the progress of improvement in the arts and science.[21]

Evans wanted data and principles to aid him in his work of invention. Neither was available in his time, and this deficiency remained a hallmark of American technology for most of the nineteenth century. The Scottish civil engineer David Stevenson remarked in 1838 that "on minutely examining the most approved American steamers, I found it impossible to trace any *general* principles which seem to have served as guides for their construction." [22] In 1824, it was predicted that further improvements in the steam engine would come about through "the auxiliary efforts of the Engineer and the Philosopher." [23] Although Alonzo Potter insisted in 1841 that "the chemical laws of heat have been investigated, with equal ingenuity and success," he admitted that "the application of these laws to the process of producing and using heat has been less studied, and has by no means made the progress, which might have been anticipated,

---

19. *Ibid.*, p. 139.
20. Oliver Evans, *Oliver Evans to His Counsel, Who are Engaged in Defense of His Patent Rights* (n.p., 1816), p. 41.
21. Evans, *Abortion*, p. 139.
22. David Stevenson, *Sketch of the Civil Engineering of North America*, 2d ed. (London, 1859), p. 71.
23. A. B. Quinby, "On Crank Motion," *American Journal of Science*, VII (1824), 322–23.

from its importance." He echoed the familiar complaint that "the greatest want, connected with the practical economy of heat, is that of *fixed principles.*"[24] As late as 1852, the *Scientific American* lamented that "there is no precise theory for the steamboiler, from which all the necessary data can be derived for the construction of a boiler of any given evaporating power."[25] To the technologist of the first half of the 1800s, science often meant fixed principles: he felt a need for them and sometimes was forced to discover them for himself.

One important factor which helped determine the extent of useful intercourse between science and industry was simply the number and distribution of scientists themselves. The small body of colonial savants clustered around the seaports of Boston, New York, and Philadelphia over the years became a legion of naturalists and philosophers inhabiting many smaller American towns. Many of these scientists were receptive to inquiries from manufacturers with technical problems. The American Philosophical Society received a number of calls for expert advice and encouragement. The chemist Thomas Cooper edited his *Emporium of Arts and Sciences* in the second decade of the century in an effort to bring those two interests together. The movement for mechanics' institutes brought men of learning and labor together; sometimes the result was mutually beneficial.

The multiplying number of scientists available on college campuses also played an important role. George Corliss, the Rhode Island steam engine inventor, is said to have received advice from Alexis Caswell, professor of mathematics and natural philosophy at Brown University.[26] In Cincinnati, in the late 1830s, a bell founder discovered that a five-ton shipment of copper pigs from Liverpool contained an impurity which resisted all his efforts to extract it, and he was "exposed to some degree of ridicule and threatened loss" on

---

24. Potter, *Principles,* pp. 85–86.
25. *Scientific American,* VIII (December 25, 1852), 115. For more on the steam engine, see Carroll Pursell, *Early Stationary Steam Engines in America: A Study in the Migration of a Technology* (Washington, D.C., 1969), pp. 113–28.
26. Robert S. Holding, "George H. Corliss of Providence, Inventor," *Rhode Island History,* V (January, 1946), 7.

this account. Charles Cist, a leading citizen of Cincinnati, proudly recorded that

> The skill and science of our professor of chemistry in the Medical College of Ohio extricated him, however, from the difficulty; the foreign substance was discharged, a fine body of pure metal run off, and more than a thousand dollars profit resulted from the adventure. I cite this as a proof of the value of men of science, too often, in communities, considered mere theorists.[27]

Treadwell at Harvard and other faculty members elsewhere exercised their ingenuity on practical problems of their own choosing: "[Treadwell's] lectures required but a part of his time, and left him free to engage in other pursuits, and he directed his attention to the making of cannon of greater strength."[28] Besides doing weapons research, he found time to establish the Cambridge Scientific Club, serve as recording secretary and later as vice-president of the American Academy of Arts and Sciences, and sit on blue-ribbon commissions to study the problems of furnishing the city of Boston with water and to investigate the Massachusetts state standards of weights and measures.[29]

THE MINERAL INDUSTRY

Another exemplary career was that of J. P. Lesley. After graduating from the University of Pennsylvania in 1838, he worked under Henry D. Rogers on the state geological survey. He later entered the ministry, but left it in 1852 and once again took up geological work. In 1856, he published *A Manual of Coal and Its Topography,* and in that same year became secretary of the American Iron Association. He also worked as a private consultant, and in 1857 his office stationery carried the following letterhead: "Geology and Topography. Geological and other Maps constructed; Surveys of Coal Lands

---

27. Charles Cist, *Cincinnati in 1841: Its Early Annals and Future Prospects* (Cincinnati, Ohio, 1841), p. 243.

28. Morrill Wyman, "Daniel Treadwell, Inventor," *Atlantic Monthly,* XXXII (October, 1873), 476–77.

29. *Dictionary of American Biography* (hereafter DAB), s.v. "Treadwell, Daniel."

made; Mineral Deposits examined; Geological Opinions given to guide purchasers, and Reports made to Owners and Agents. Orders for elaborate Topographical Surveys from Rail-road and other companies, will be executed in scientific principles, and in the highest style of the art."[30] Two years later he joined the faculty of the University of Pennsylvania, was made dean of the Science Department in 1872, and dean of the new Towne Scientific School in 1875. He was librarian, secretary, and vice-president of the American Philosophical Society, and a charter member of the National Academy of Sciences. During all this time, he continued his consulting activities, traveling in 1863 to Europe for the Pennsylvania Railroad to study the Bessemer steel process. Somehow he also found time to serve as state geologist, direct the second Pennsylvania geological survey, and edit for four years a weekly newspaper, *The United States Railroad and Mining Register*.[31]

Lesley's story was probably duplicated many times over in the entrepreneurial atmosphere of mid-century American science. Scientists on the Yale faculty seemed to spend as much time away from campus in the post-Civil War era as their descendants do today. In 1864, the Yale chemist Benjamin Silliman earned a handsome $50,000 for a nine-month tour of the West doing consulting work. Both Josiah D. Whitney and William H. Brewer worked during these same years for the California state geological survey, which Whitney headed.[32] It would be fascinating to know exactly how many academic scientists in the second half of the century supplemented their collegiate incomes with retainers and consulting fees of various kinds.

Since geologists were the most numerous of scientific specialists in the United States of the nineteenth century, and since work in geology accounted for some of the most distinguished contributions

---

30. Joseph Lesley, Jr., to James Hall, June 15, 1857, George P. Merrill Papers, Library of Congress, Washington, D.C.

31. *DAB*, s.v. "Lesley, Peter."

32. Gerald T. White, "The Case of the Salted Sample: A California Oil Industry Skeleton," *Pacific Historical Review*, XXXV (May, 1966), 153–84. See also *idem, Scientists in Conflict: The Beginnings of the Oil Industry in California* (San Marino, Calif., 1968).

by Americans to the general body of science, it is no surprise that the mining industry was one of the first to feel the quickening influence of increasing and maturing scientific activity. A man like James Curtis Booth moved easily and usefully between science and industry, chemistry and geology. He graduated from the University of Pennsylvania, after having studied chemistry with Robert Hare and William Keating. After another year's study at Rensselaer Polytechnic Institute, he went to Germany where he visited chemical plants and studied with Wöhler and Magnus. After his return to Philadelphia in 1836, he established a private laboratory where he did consulting work and instructed students in laboratory technique. He taught at the Franklin Institute, Philadelphia Central High School, and the University of Pennsylvania, and served with the state geological survey before becoming director of the survey of Delaware. He was active in many scientific groups and president of the American Chemical Society. Throughout his life, he advised on geological and metallurgical matters in industry.[33]

The center of geological activity, of course, lay with the United States Geological Survey after its establishment in 1879. In the field, as well as in the Washington headquarters, the USGS formed intimate bonds with the mining industry. Rodman Paul has described its operations in the Colorado mining region. Facing novel forms of mineral deposit, the miners, in Paul's words, "needed *scientific* solutions. That is to say, they needed an understanding and a set of processes that could come only from chemistry, metallurgy, and geology." One of the first scientists at the scene was Nathaniel P. Hill, professor of applied chemistry at Brown University, who was sent out in the mid-1860s by New England capitalists to save their investments. After visiting Wales and Germany, in addition to Colorado, Hill imported a Welsh foreman and in 1868 succeeded in setting up a smelter capable of handling the local ores.

Smelters were built in Leadville in the late seventies by men with training in European mining schools, and during 1879–81 Samuel F. Emmons of the USGS made an important report on the area.

---

33. *DAB*, s.v. "Booth, James Curtis."

243

In 1882, he helped found the Colorado Scientific Society in Denver, where government geologists mixed with metallurgists, assayers, chemists, mining engineers, mining managers, and other geologists. When the Cripple Creek gold strike of 1890–91 presented novel problems, both the local society and the USGS investigated and diagnosed the condition, and made suggestions for economical exploitation—all quickly enough to prove commercially useful. The episode was a superb example of entrepreneurial activity in a pluralistic setting.[34] The growth of petroleum as an important part of the mining industry reinforced the demand for scientific expertise, as the careers of Lesley, Silliman, Herman Frasch, and others testify.[35]

Besides independent consultants like Frasch, government scientists like Emmons, and university professors like Hill, the mining industry was also able to hire a growing body of engineers trained in both the theory and practice of mining. It was estimated that in 1893 some 6,000 technical people were employed by the American mining industry and that there were four positions open for every mining engineer graduated from an American college. Between 1819 and 1865, at least seventy-six Americans had studied at the great mining school at Freiberg, and, in the years after the war, curricula in mining engineering spread to many of the land-grant and private colleges of this country.[36] The American Institute of Mining Engineers became the first engineering specialty to break away from the ecumenical American Society of Civil Engineers.

By the 1890s, however, mining engineers were already eclipsed by mechanical and electrical engineers in both number and demand. In 1812, one observer had noted that "there appears in the United States a redundancy of young men, *collegiately educated in the arts and sciences,* and of the classics, which usually apply to divinity,

---

34. Rodman Wilson Paul, "Colorado as a Pioneer of Science in the Mining West," *Mississippi Valley Historical Review,* XLVII (June, 1960), 34–50.

35. Herman Frasch, "Address of Acceptance," *Journal of Industrial and Engineering Chemistry,* IV (February, 1912), 134–40.

36. Samuel P. Christy, "The Growth of American Mining Schools and Their Relation to the Mining Industry," *Transactions of the American Institute of Mining Engineers,* XXIII (1893), 444–65. See also Clark Spence, *Mining Engineers and the American West: The Lace-Boot Brigade, 1849–1933* (New Haven, Conn., 1970).

medicine, law and commerce." Such young men should rather apply to industry. "It is a truth," he continued, "that a knowledge of the arts and sciences, is extremely valuable as a *preparative* and *an accompanyment* to an intelligent apprenticeship to manufactures. Mechanism, chemistry, metallurgy, hydraulics, geometry, mensuration, pneumatics, and mechanic powers, natural history, . . . are very useful, and almost necessary in the great manufacturing establishments, which enrich Europe." [37]

THE ENTREPRENEURIAL SPIRIT

Ben-David points out that the United States has historically given practical collegiate training to a comparatively large proportion of its young people.[38] In the 1800s, a host of practical curricula, approved by college presidents seeking to expand the markets for their products, turned out increasing numbers of young people to whom science was neither a chaste mistress nor an idle fancy, but rather a useful tool. "With such a student," wrote M.I.T. President Francis A. Walker, "the useful applications of science distinctly add to the educational value of scientific study, inasmuch as they give a more direct object to his efforts and exertions, and heighten the pleasure he feels at each step of his scholarly progress." [39]

The engineer, among all these graduates, was the best qualified to join theory with practice; his education and experience gave him that general knowledge of science and specific knowledge of industry which maximized the opportunity for useful innovation. The engineer-entrepreneur has become a familiar figure in the history of American technology, but he also deserves a place in the history of American science.[40] In 1891 Cyrus F. Brackett, Henry Professor of Physics at Princeton, put the matter in terms that Americans of the Gilded Age could understand: "The present offers to the student the ac-

---

37. *Niles' Weekly Register*, II, supp. (March 21, 1812), 53.
38. Ben-David, *Fundamental Research*, p. 37.
39. Francis A. Walker, "The Technical School and the University," *Technology Quarterly*, VI (1893), 228.
40. W. Paul Strassmann, *Risk and Technological Innovation: American Manufacturing Methods during the Nineteenth Century* (Ithaca, N.Y., 1959), p. 9.

cumulated treasures of knowledge and the hope of scientific distinction as well as that of pecuniary reward."[41]

The hope of pecuniary reward was an important constraint, as well as stimulant, for the normal pursuit of science had certain weaknesses when viewed from the perspective of application. Science tended to move ahead in narrow and widely separated salients so that many of those principles which Oliver Evans sought at the beginning of the century were still lacking at the end. F. W. Clarke, chief chemist of the USGS, attempted in 1891 to account for this incomplete knowledge. He explained:

> Apart from the vastness of the field to be explored, itself a sufficient excuse for ignorance, the more obvious deficiencies are due to excessive individualism in research. Thousands of earnest men are working independently, with insufficient reference to one another, each attacking that corner of the unknown which most attracts his fancy. All are ambitious to accomplish great results, each one hopes to make some discovery of signal importance; and so the drier and less attractive details of investigation are oftentimes neglected. The field is cut up in many fields, between which the ground is uncultivated, and there no harvest is gathered.[42]

The sociology of engineering was strikingly different. Engineers gained recognition not by advancing science but by applying it. In shipbuilding, petroleum refining, mining, chemical manufacture —in one industry after another—they moved into positions of usefulness and responsibility. The electric motor brought them to many factories, and the Bessemer process introduced them to the iron industry.[43] The Pennsylvania Railroad, one of the largest and most complex corporations of the nineteenth century, was presided over

---

41. Cyrus F. Brackett, "The Effect of Invention Upon the Progress of Electrical Science," *Proceedings of the Celebration of the American Patent System* (Washington, D.C., 1891), p. 291.

42. F. W. Clarke, "The Relations of Abstract Scientific Research on Practical Invention, with Special Reference to Chemistry and Physics," *ibid.*, p. 309.

43. John Fritz, "The Progress in the Manufacture of Iron and Steel in America, and the Relations of the Engineer to It," *Transactions of the American Society of Mechanical Engineers*, XVIII (1896–97), 39–69; Carroll W. Pursell, Jr., "Durfee's 'Pothecary Shop: The Chemical Laboratory at Wyandotte, 1863," *Detroit Historical Society Bulletin*, XXII (January, 1966), 4–9.

by engineers for ninety-six of the first one hundred years of its existence, and its employment of Charles B. Dudley to test materials and standardize specifications was one of the landmarks of the increasing usefulness of science to industry.[44] The origins of the movement to establish federally sponsored engineering experiment stations were said to lie with "Southern colleges, and members of the engineering profession who desire to see the encouragement of Southern industries through scientific method."[45]

Even more important than the particular examples of applied science attributable to engineers was the impetus they gave to what Robert Wiebe has called a search for order.[46] In a recent study of American engineering during these years, Raymond Merritt has noted that "the new professionals came armed with technical training, administrative experience, well-prepared statistical studies, and operational plans. They also displayed a strong sense of personal integrity and public responsibility." Dedicated to public service, knowledge, and efficiency, equipped with expert training, professional standards of conduct, and a cosmopolitan outlook, these engineers worked "toward an alternative to the entrepreneur's concern with independent initiative, often diverting a rising corporation toward the goal of social utility and making it a tool of noncompetitive consolidation."[47]

Whether serving international corporations or local municipal governments, engineers brought with them all the stigmata of bureaucratic expertise: rational organization and control through the exercise of special knowledge, disciplined by professional standards. Engineering virtues of hierarchy, discipline, rationality, utility, efficiency, and uniformity were generalized to become the guiding principles of corporate America, and eventually of government as well. With these values, the engineer was the sworn enemy of provincialism, eccentricity, individuality, tradition, dishonesty, and

---

44. *Memorial Volume Commemorative of the Life and Life-Work of Charles Benjamin Dudley, Ph.D.* (Philadelphia, Pa., 1911).

45. *Science,* III (January 3, 1896), 20.

46. Robert H. Wiebe, *The Search for Order, 1877–1920* (New York, 1967).

47. Raymond H. Merritt, *Engineering in American Society, 1850–75* (Lexington, Ky., 1969), p. 7.

waste. In short, science became the standard, not merely the handmaiden, of progress.

The story of the relationships between science and industry in nineteenth-century America, therefore, is much more than a groping toward institutionalization of industrial research in the modern corporate laboratory. It is the realization of the rich diversity of science in America during these years—the presence of science in a host of different institutions, its dissemination by disparate agencies, and its practice by thousands of college teachers, government bureaucrats, private consultants, amateur naturalists, and employees of manufacturing firms. The very plurality and variety of forms led to a strengthening of the entrepreneurial spirit that was a major characteristic of the American scientific system. One of the fruits of that spirit was a common and practical application of science in industry.

The Promise of the Future: Technical Education

Cotton Mather, the eighteenth-century divine, once called theology and medicine the "angelic conjunction." For Americans in the nineteenth century, the heaven-sanctioned combination was technology and democracy. The Industrial Revolution and the American Revolution seemed fortuitously and inextricably joined as the parents of a politically creative and mechanically ingenious new nation.[1] Technical education in nineteenth-century America seemed marked by the same happy association. One had only to create opportunities for the acquisition of useful knowledge, and the door to political equality, economic security, and social justice—the promise of the future—lay open for all.

Useful knowledge, a phrase that came easily to the lips of Americans during the last century—especially in the early years—carried a web of connotations. It was used interchangeably with the word *practical;* both terms meant the opposite of abstruse and theoretical knowledge, which only the rich or well born had traditionally enjoyed. *Useful* meant democratic and was what Americans had in mind when they talked about technical education; it was almost always charged with a reform mission.[2] A properly functioning,

---

1. See Hugo A. Meier, "Technology and Democracy, 1800–1860," *Mississippi Valley Historical Review,* XLIII (March, 1957), 618–40. As Brooke Hindle pointed out in *Technology in Early America: Needs and Opportunities for Study* (Chapel Hill, N.C., 1966), the history of technical education in America remains to be written. His bibliographical suggestions, plus those in Eugene S. Ferguson's *Bibliography of the History of Technology* (Cambridge, Mass., 1968), are the best places to begin.

2. See David M. Potter, *People of Plenty: Economic Abundance and the American Character* (Chicago, 1958), pp. 128–41.

harmonious society depended on the balance struck between political ideology, educational opportunity, and the nation's material advance. By opening avenues of self-determination, technical education was to be democracy's mainspring; by eliminating class distinctions, it was to be society's balance wheel. And since useful knowledge was the key to America's natural abundance, the virtue of the political system would be proved to a skeptical Old World by the wealth which would surely follow. This appealing vision was maintained, at least on a rhetorical level, with remarkable persistence throughout the century.

As much as it might appear that Americans had a consistent approach to technical education, the agencies that were created often reflected a divergent reality. How this new kind of learning would be related to existing educational structures, what should be taught, what teaching methods should be used, for whom the training would be designed, and by whom it would be supported—all these were complex problems which required solutions. (Some of them have not been resolved yet.) Institutional development was uncertain and uneven. Despite diversity, confusion, and tension, there was a general pattern of development in technical education and most urban centers went through essentially the same stages of growth. Philadelphia is a particularly good example for study because of the number of educational experiments tried there and because, as a center of manufacturing industry, it placed special emphasis on technical training.

TECHNOLOGY VERSUS THE CLASSICS

At the beginning of the century, organized instruction in science and its applications was unavailable in Philadelphia. The military academy at West Point was the only formal educational agency which existed before the 1820s to provide Americans with technical education.[3] One could gain a certain amount of knowledge through

---

3. Sidney Forman, *West Point: A History of the United States Military Academy* (New York, 1950). See also Daniel H. Calhoun, *The American Civil Engineer* (Cambridge, Mass., 1960); and Forest G. Hill, *Roads, Rails and Waterways: The Army Engineers and Early Transportation* (Norman, Okla., 1957).

informal channels. In Philadelphia, as elsewhere, itinerant lecturers regularly visited the city and discoursed on chemistry, electricity, and the steam engine. Urban machine shops provided another kind of educational opportunity for inquiring young men with a bent for mechanical knowledge.[4] But in the classroom, instruction continued to focus on the traditional, classics-dominated curriculum until the 1820s, when economic prosperity and a wave of egalitarian sentiment stimulated the creation of a variety of organizations interested in technical education. Most of these associations—apprentices' libraries, village lyceums, and mechanics' institutes—were brought into existence by an unsatisfied demand for useful knowledge, and they were nourished by the conviction that by studying the practical uses of science, a man could improve himself and his chances for the future.[5]

Of all the organizations formed during that remarkable decade, none achieved more immediate fame than the Franklin Institute of the State of Pennsylvania. The Institute, like other societies at the time, was founded to implement the promises of the Declaration of Independence by providing Philadelphia's citizenry with "the unspeakable blessings of education." Its leaders claimed as their central goal the establishment of "that equality so particularly recognised in our Bill of Rights."[6] For want of other means, these informal associations seemed the best expression of democratic ideals and the best agencies to spread useful knowledge. But however similar their aims, their efforts lacked precise definition of whom they would educate or what methods they would use. All such organizations, whether the Rensselaer school in New York, the Gardiner Lyceum in Maine, or the Franklin Institute in Pennsylvania, had eventually to face the challenge of casting their reforms into more rigorous educational programs or of accepting a limited, relatively ineffective role.

The Franklin Institute's initial attempts at technical education

---

4. See Eugene S. Ferguson, *Early Engineering Reminiscences* [1815–40] *of George Escol Sellers* (Washington, D.C., 1965), p. 16.

5. See Carl Bode, *The American Lyceum: Town Meeting of the Mind* (New York, 1956).

6. The Memorial of the Officers and Board of Managers of the Franklin Institute of Pennsylvania for the Promotion of the Mechanical Arts . . . 26th February, 1824, Franklin Institute Archives, Philadelphia, Pa.

were characteristic of most voluntary associations. At first, there were evening lectures on miscellaneous subjects to a mixed audience. In the second season of operation, a more formal program of instruction was inaugurated. Ten-week courses in specific subjects were given by paid faculty. That form of organization, however, proved unsatisfactory. The lectures were too abstruse for the audience. Yet to lower the intellectual content seemed inconsistent with the Institute's educational ambitions. While the Institute's Committee on Instruction pondered the problem, Peter A. Browne, a local attorney, one of the society's founders and a dabbler in science, presented the Board of Managers with a novel plan. He argued that real knowledge of the sciences was available only in colleges and universities, where entrance was contingent on a knowledge of Latin and Greek. Since the "operative members of society," the working classes, could not spare the time to learn those languages, "the system tends to shut the door of science against a numerous class of the most useful citizens." Browne therefore urged that a college be attached to the Institute which would teach all the subjects that were necessary to create the "Scientific Mechanic or Manufacturer," at a cost low enough to make the education available to all.[7] Courses in Greek and Latin were specifically excluded from the curriculum.

Within the Board of Managers, reaction to Browne's idea was ambivalent and triggered an alternative plan from the Committee on Instruction, which proposed that the Institute establish a high school department to teach a mixture of science, liberal arts courses, Latin, Greek, and possibly modern languages.[8] In the ensuing months debate over the two ideas provided the basis for determining the Institute's position on technical education. Was the curriculum to be directly practical or academic in content? Was the new department to be a genuine educational opportunity for the working-class poor or a mechanism for the advancement of those who might

7. "Minutes of the Board of Managers of the Franklin Institute," February 2, 1826, Franklin Institute Archives; [Peter A. Browne], *A Plan of a College to be Attached to the Franklin Institute* (n.p., n.d.). William Stanton, *The Leopard's Spots: Scientific Attitudes toward Race in America, 1815–1859* (Chicago, 1966), pp. 149–54.

8. Committee of Instruction, Draft of a Report by Mr. Merrick, March, 1826, Franklin Institute Archives.

be described as middle class in occupation and attitude? Was technical education to be a reform, or did it have other objectives? As the alternative proposals took shape, argument over the classical languages collected several issues into a single debate.

The essence of Browne's approach was that classical study was a waste of time for boys who would be apprenticed to trades and crafts. In the period between the age of ten, when they left common school, and the age of fifteen, when they began their apprenticeships, they did not have time to learn the languages well, nor had they any use for the knowledge they did gain. It was ridiculous, Browne argued, to claim that the University of Pennsylvania's science courses were easily available to those who wished additional knowledge; in a city of 130,000, the university's enrollment was sixty-four students. The only answer was a scientific and technical college which would "embrace every branch of instruction required for the agriculturist, the mechanic or manufacturer, the architect, the civil engineer, the merchant, and other man of business." [9]

The position of the Committee on Instruction was elaborated by Samuel Vaughan Merrick, another of the Institute's founders and a representative of the city's emerging industrial elite. To eliminate all study of Latin and Greek meant barring admission to any who might wish to continue on to the university. Many artisans felt the same, Merrick claimed:

> Amongst enlighten'd Mechanics I find this sort of feeling existing with respect to Browne's college: "Let them establish a college in which we can have the option of learning the languages if we please, but do not set up the doctrine that we have no right to the opportunity, we will admit of no such invidious distinction." [10]

The Franklin Institute's high school, Merrick argued, should provide a thorough education for the young man who would be a

---

9. *To the Citizens of the City and County of Philadelphia* (Philadelphia, Pa., 1826). See also Mathew Carey, *Reflexions on the Proposed Plan for Establishing a College in Philadelphia* (Philadelphia, Pa., 1826); John Sanderson, *Remarks on the Plan of a College (About to be established in this City)* (Philadelphia, Pa., 1826); and *Address to the Trustees of the Polytechnic and Scientific College to be established in Philadelphia* (Philadelphia, Pa., 1826).

10. S. V. M[errick] to (?), March 16, 1826, High School File, Franklin Institute Archives.

mechanic, but it should also give him the opportunity for "the higher branches of education, which are only attainable in the colleges."[11] He saw the high school as a direct link between the city's common schools and the University of Pennsylvania and as a part of Philadelphia's established educational system. To strengthen the connection even further, he proposed merit scholarships to the university for the best students of the high school. Since Merrick's colleague on the Committee on Instruction, Robert Patterson, was vice-provost of the university, there was no real impediment from that quarter. But Merrick was aware of sympathy in the Board of Managers for directly practical knowledge, and thus he suggested clouding the issue. As he pointed out in a private communication, "The uselessness of the dead languages to operatives is a popular theme in our board at present," and he counseled keeping "the main object in the Back ground," emphasizing instead "the necessity of a mechanical education to meet the demands of the age." The classics could later be slipped in through the back door.[12]

The Institute's Committee on Instruction never argued, as Browne did for his college, that the high school should provide education for those who had been denied it. They proposed instead to give those of moderate means the educational opportunities which had previously been open only to the wealthy. It was more democratic, the committee claimed, to keep the avenues of social and economic advancement open than to fix class lines by separate educational systems. To deny anyone the advantages of a liberal education was to deny professional distinction, social esteem, or political preferment.[13] These were the ambitions to be served by the high school. With this reasoning, the Franklin Institute rejected Browne's proposal and adopted instead the plan for a high school, which was opened in 1826.

In operation, the school mirrored the aspirations which had given

---

11. *Ibid.*
12. *Ibid.*
13. *Address of the Committee on Instruction of the Franklin Institute of Pennsylvania, on the subject of the High School Department attached to that Institution* (Philadelphia, Pa., 1826). The question of whether or not the classics would interfere with political ambition was discussed in most of the pamphlets printed during the dispute.

it life. As one student later recalled, it was "well patronized by the best families of the city."[14] The course of studies included those subjects, the committee said, which would be "universally selected by the enlightened parent, whose wealth enables him to make a choice for his son." If the student took the full three-year course, he had three years of Greek and Latin, three years of mathematics and French, two years of Spanish and drawing, plus courses in history, geography, political economy, astronomy, natural philosophy, natural history, chemistry, and bookkeeping. Additional training in science was provided by the Institute's regular evening lectures.[15] It was exactly the curriculum needed for the young man who might go on to the university or, more likely, into the family's manufacturing establishment.

Philadelphians were not unique in their educational efforts or in their aims. The New York Mechanic and Scientific Association, mainly through the energies of John Griscom, had opened a high school in 1825, one year before the Philadelphia school opened. It also drew its students from "the best families," and provided similar courses for the same reasons.[16] Both schools were the new kinds of voluntary educational associations the governor of Connecticut had in mind when he addressed the state legislature in 1825:

> By the common progress of science in our country, institutions are gradually forming, which are designed to promote every branch of useful knowledge, with appropriate applications to the minds of young men of the principles of mathematics, chemistry, geology, mineralogy, botany, zoology, and natural philosophy. By a knowledge of these sciences, they become intelligent agriculturalists, machinists, manufacturers, architects, and civil and military engineers.[17]

These informal organizations established some important precedents for technical education in America. Philadelphia's quarrel

---

14. Elizabeth B. Pharo, *Reminiscences of William Hasell Wilson, 1811–1902* (Philadelphia, Pa., 1937), p. 40.
15. *Address of the Committee on Instruction of the Franklin Institute,* pp. 3–4.
16. John Griscom, *Monitorial Instruction: An Address, Pronounced at the Opening of the New-York High-School* (New York, 1825), p. 213.
17. As quoted in Griscom, *ibid.,* p. 26.

over Latin and Greek was part of a larger process of determining what technical education should be, where it should be taught, and who should be instructed. Despite their disagreements, Browne and Merrick shared certain implicit assumptions. The first was that contemporary generations were too wedded to existing techniques to be aided significantly by organized technical instruction; both men fastened their hopes on "the rising generation." Technical education would not take the form of adult education; it contained no concept of retraining or of skill improvement. The two men shared the belief that the kind of instruction which they had in mind necessitated special educational structures. Existing institutions did not answer the need. When the Institute's Board of Managers rejected Peter Browne's college, they also rejected the idea that a high school should be vocationally oriented. It should teach those sciences upon which engineering skills might be based, but not the kind of knowledge that apprentices would employ as they entered their trades. Without the support of Philadelphia's industrial community, or at least its leadership, Browne's idea quickly died. With it died any viable link between reform zeal and technical education. The Franklin Institute's high school did not alter the educational circumstances of the working-class poor. Because classes were conducted during the day, its facilities were unavailable to apprentices.[18]

The advent of tax-supported high schools did not change anything. They tended to be patterned after private schools and were frequently directed by the same people. Public funds only made it easier to realize the same objectives. When the Pennsylvania legislature provided support for high schools in 1836, the designers of the Institute's educational program played a major role in shaping the city's new Central High School.[19] Alexander Dallas Bache was

---

18. The principal of the school was Walter R. Johnson, who later went on to some fame as a scientific investigator. While the school was initially successful, a disagreement between Johnson and the Institute over finances led to its closing three years later. See George E. Pettingill, "Walter Rogers Johnson," *Journal of the Franklin Institute,* CCL (August, 1950), 93–113.

19. The Institute, with Bache as chairman of the Committee on Instruction, attempted without success to carry on a school in the years between the closing of Johnson's school in 1831 and the establishment of Central High.

appointed Central High's first principal when the school opened in 1838. With his several years of experience as chairman of the Institute's Committee on Instruction, a faculty comprised largely of Institute professors, and his own educational background at West Point, Bache fashioned a school which blended emphasis on the sciences, their industrial applications, and the traditional college-preparatory studies.[20] His approach was perfectly tailored to the needs of the school's constituency. Urban centers, the loci of the most active support for public education, were just beginning to react to the emergence of industrial capitalism. And as Michael Katz has pointed out in his study of education in Massachusetts, high schools found their champions not in the ranks of the city's poor, who resented the burden of higher taxes, but among those more prosperous members of society who viewed education as a bulwark to their own position.[21]

What ambitious middle-class supporters of technical education also saw was the need for increasingly sophisticated knowledge in order to most rationally exploit America's economic potential. The nation's material advance was always part of the litany which linked democratic ideology and useful education. But the sharp sectional competition in the decades before the Civil War stimulated the idea that specialized skills were the best guarantee for economic survival. Railroad commissioners might damn the West Point engineer for his intellectual snobbery, but technical schools of the time tried to imitate the standards of his education.[22] Americans talked practi-

---

20. George Davidson, who worked with Bache in the Coast Survey, later remembered, "He told me his ambition had been to make the Central High School equal to West Point in all points pertaining to a thoroughly practical education, to fit a man for the duties of his professional or civil career." Franklin Spencer Edmonds, *History of the Central High School of Philadelphia* (Philadelphia, Pa., 1902), pp. 78–79.

21. Michael B. Katz, *The Irony of Early School Reform: Educational Innovation in Mid-Nineteenth-Century Massachusetts* (Cambridge, Mass., 1968).

22. Calhoun, *The American Civil Engineer*, p. 137. At the University of Michigan, for example, Henry Tappan claimed, "We followed as a model the military academy at West Point, and succeeded, after much effort, in securing two professors educated there to conduct it. Our aim was to organize a scientific school of the highest character." H. P. Tappan, *Review by Rev. H. P. Tappan of his connection with the University of Michigan* (Detroit, Mich., 1864).

cality, but they created institutions which emphasized theoretical studies.

THE TECHNICAL UNIVERSITY

Once high schools were fitted into the educational establishment, with an academic rather than a vocational mission, advanced technical instruction found its home in the university or in university-level institutes. In Philadelphia, this second tendency followed directly on the heels of the first. The same year Central High opened, Philadelphians petitioned the legislature for funds to create a state technical university. It was to be attached to the Franklin Institute and called the School of Arts, but in every other respect it was analogous to the agricultural and mechanical colleges later established by the Morrill Act of 1863.

The principal objective of the proposed school, "one vast *University*" as its promoters described it, was to unite science and practice in a new combination. Previous efforts to teach the artisan those scientific principles which underlay his craft had generally been limited to demonstrating the action of simple mechanical powers and the chemistry involved in elementary industrial processes. Or, at least, the artisan seemed not to have absorbed much more than that. One always hoped the first taste of science would lead to another, but the school's advocates suggested there was no longer time to hope. In the iron industry, for example, they claimed that Pennsylvania was economically threatened by technological advances in New York and New Jersey. Science alone could replace outmoded practices and secure a trade which the commonwealth should "have monopolized."[23] Similarly, the coal mining industry—if the state properly supported the School of Arts—could be made the great economic asset which cotton production was for the South. What was needed was a technical school which would incorporate an experimental farm, operational workshops, and an analytical laboratory. In these adjuncts to the classroom, theory could be applied directly

---

23. *Memorial of a Committee appointed at the town meeting of the citizens of the city and County of Philadelphia, Held January 4, 1838, Praying for the Establishment of a School of Arts* (Philadelphia, Pa., 1838), p. 6.

to industrial production, resource development, and agricultural improvement.

Instruction in the school was to be organized into six departments: mechanical science; chemical arts; mathematics, geology, mineralogy, and mining; civil engineering and architecture; agriculture and rural economy. Within the departments, teaching, laboratory training, and field experience were all structured so that the best technological solutions were defined as those most economically feasible. The student in the mining course, for instance, would take lessons in the field and employ the science of geology to locate and assess mineral resources. He would also learn the techniques of mine construction, extraction processes, ore dressing and handling, and the surveying and plotting of mine fields. Geology and mineralogy would form the theoretical base of his instruction; the practical side would include all the techniques necessary for converting a natural resource into a commercial product. Rational technology was the key to obtaining the earth's wealth, "abundantly and at least expense." [24]

In such a single-minded educational plan, there was neither time for nor point to the traditional liberal studies. But because theoretical science played a key role in their proposal, the Philadelphians conceived a school intellectually more sophisticated than a vocational institution. Its graduates, therefore, deserved some special place in the community, even if they lacked the social polish that might normally come from a higher education. The answer was one that has been attractive to engineers ever since. The proposed program was to be a "professional education," similar to the specialized courses required of university-trained attorneys and physicians. [25] Since it was an equivalent education, the graduates would presumably enjoy the same social esteem as those in the other learned professions.

But 1838 was fiscally a bad year for Pennsylvania; the legislature was unsympathetic to any further drain on its funds, and the idea

---

24. For the Establishment of a School of Arts: Memorial of the Franklin Institute, of the State of Pennsylvania, for the Promotion of the Mechanic Arts, to the Legislature of Pennsylvania (Philadelphia, Pa., 1837), p. 8.

25. Ibid., p. 11. See also American Railroad Journal and Advocate of Internal Improvements, VI (January 14, 1837), p. 4.

was never carried into effect.[26] Even though it was stillborn, the proposal reflects another important plateau for American technical education. For the remainder of the century, when Americans talked about technical education, they meant advanced education. Their restricted definition always surprised Europeans, who took the term to encompass a wide variety of schooling, from purely vocational instruction at an elementary level to highly theoretical study at a university level. Because Americans came to think of technical education as higher education, they also came to think of it in terms of a specialized curriculum for specific occupational groups. However, since technical schools were still in the process of defining clearly their form, function, and clientele, funding was the central problem.

Many schools shifted from one course of studies to another, in the hope of catching a fresh breath of support. The Rensselaer school, for example, came to specialize in engineering only through a series of educational experiments after Stephen Van Rensselaer had withdrawn his financial aid.[27] But as industrial development took on a national character, and as Americans became aware of Old World advances in technical education, form, function, and clientele emerged. The European polytechnic institute, which was state supported and highly organized, posed a threat which Americans could only interpret—with a mixture of concern and envy—in economic terms. By the early 1850s the European model bestirred Americans in and out of the universities, and, with their desire to meet the challenge, they shed a bit more of the ideological baggage with which they had started.

The influence of the polytechnic idea in America was remarkable for the simultaneity of its impact. Both Harvard and Yale inaugurated

---

26. Another appeal to the legislature was made in the following year, but it was rejected, too. For a time the Franklin Institute thought of conducting the school itself, but Alexander Dallas Bache, who had just returned to Philadelphia from a tour of European educational institutions, advised against it and the idea was finally dropped. "Report of the Committee on Instruction," September 19, 1839, Franklin Institute Archives; and A. D. Bache, *Report on Education in Europe, to the Trustees of the Girard College for Orphans* (Philadelphia, Pa., 1839).

27. Samuel Rezneck, *Education for a Technological Society: A Sesquicentennial History of Rensselaer Polytechnic Institute* (Troy, N.Y., 1968), pp. 42–43.

departments for teaching the applications of science in 1847, the same year William Barton Rogers published a plan which ultimately led to the foundation of the Massachusetts Institute of Technology.[28] The University of Pennsylvania also took its first tentative steps toward a separate department of science in 1847; Rensselaer reorganized along polytechnic lines two years later; and within the next decade new departments or technical institutes were spread throughout the country.[29] The reason for America's wide imitation lies in the fact that the polytechnic reinforced trends in technical education which were already beginning to take shape. By the 1840s, Americans had concluded that industrial competition required specialized educational agencies, that the instruction should be at the university level, and that it should be pointed toward the training of a new, technical-managerial class. European models gave those thoughts a sharper definition and clearer objectives.

The crusade of Alfred Kennedy illustrates how and why Americans adopted the new form. Kennedy was trained as a chemist and physician in Philadelphia, served for a time as professor of chemistry at the Philadelphia College of Medicine, and lectured on the same subject in the Franklin Institute. In 1849, he opened a laboratory of industrial chemistry in the city, and then in the following year toured the universities and laboratories of Great Britain and the Continent. He studied chemistry in Paris and with Liebig, Ordmann, and Lehman in Germany, returned to Philadelphia late in 1852, and next spring received a charter for the Polytechnic College of the State of Pennsylvania.[30]

One of Kennedy's central objectives was to provide for the educational needs of a new occupational class. No special agency existed to train those destined for engineering and advanced manufacturing and agricultural pursuits. His school, therefore, occupied "a new position among educational institutions." Unlike West Point, it was

28. William B. Rogers, *A Plan for a Polytechnic School in Boston* (Boston, Mass., 1847). See also *idem, Objects and Plan of an Institute of Technology* (Boston, Mass., 1861).

29. Rezneck, *Education for a Technological Society,* pp. 86–92. See also Earle D. Ross, *Democracy's College* (Ames, Iowa, 1942), chap. 2.

30. *Historical Record of the Polytechnic College of the State of Pennsylvania* (Philadelphia, Pa. 1890), pp. 3–5.

not a military school, nor was it "a mere literary college." Instead, students were "instructed professionally, preparatory to entrance upon active, lucrative and honorable business." The emphasis was on professional training and the new man for whom the training was designed. "Professional Miners, Engineers, and Directors of farms and factories, have not as yet existed as a class in this country." The result was incompetence, wasted capital, and what Kennedy called "a loss of position before the art-tribunal of nations."[31]

The concept of international judgment struck a new note. To elevate America's standing before that tribunal, Kennedy consciously organized the school "on the plan of the industrial colleges of Paris and Germany." The analytical laboratory was directly modeled after those Kennedy had seen in Germany. As in the European schools, classwork and field trips had as their constant objective the advancement of industry through the applications of science. In fact, Kennedy's description of the aims of the college make clear that the tribunal of nations also bore some relationship to the marketplace. On scientifically trained mechanical engineers, for instance, fell the important obligation "of multiplying the effective industry of the nation, and by the perfection and concentration of labor-saving machinery, of enabling us in manufactures, to compete successfully with foreign countries, where labor is cheaper." America's standing in the world meant the ability to compete, and that demanded an even greater intellectual effort. Only a "systematized technical education" would provide the knowledge required by the "Agents, Superintendents, Engineers, Contractors, Directors, Presidents and Commissioners" of the great enterprises of the future.[32]

The polytechnic institute was a critical step in the evolution of technical education in America. It permanently established a highly specialized course of studies based on rigorous training in mathematics and the physical sciences. It defined the education of men

---

31. *Second Annual Announcement of the Polytechnic College, of the State of Pennsylvania* (Philadelphia, Pa., 1854), p. 6. See also, *The American Polytechnic Journal: A New Monthly Periodical devoted to Science, Mechanic Arts, and Agriculture,* I (1853), 442.
32. *Second Annual Announcement,* p. 9.

who would be managers as much as they would be technical experts, and it accepted as one of its obligations the defense of American industry against foreign competition. These objectives, explicit at the outset, still did not resolve all the issues. Advocates of the polytechnic were unable to provide a precise picture of what the institution should be. Kennedy, for example, argued that an institution established to train only mining engineers or only civil engineers would have produced "half-educated graduates." He insisted that the school's departments were interdependent and that it was their "union in a POLYTECHNIC COLLEGE" which gave the requisite breadth and depth. Other promoters of the polytechnic were equally imprecise when they attempted to define its special quality.[33] In reality, what bothered them was the polytechnic's place relative to the nation's established colleges and universities. They were aware of a certain disdain among the traditionally educated for polytechnic studies and knew that their schools did not have the prestige of other institutions. Americans had accepted the idea that technical education should be conducted at a university level, but how established colleges and universities would absorb this new branch of education had not been determined.

In contrast with technical schools, universities enjoyed both financial support and social position. In his long personal struggle to keep the Polytechnic College of Pennsylvania alive, Alfred Kennedy on occasion must have looked wistfully at the resources of the University of Pennsylvania. His own school met a need, and on that basis managed to survive a few decades. But it received little financial aid from the industries or commonwealth it wished to serve, and scant cooperation from sister institutions in the city.[34] The University seemingly reaped all those benefits, plus the heritage of Benjamin Franklin and a strong interest in science.

---

33. At Rensselaer, Benjamin F. Greene argued that the polytechnic was "materially different" than other schools, but he never succeeded in a more explicit definition and fell back on the phrase "True Polytechnic." Rezneck, *Education for a Technological Society*, p. 116.

34. See Saul Sack, *History of Higher Education in Pennsylvania* (Harrisburg, Pa., 1963), II, 481–82; and John F. Frazer to M. Newkirk, Philadelphia, April 15, 1854, Franklin Institute Archives.

The first suggestion for a separate department in the University to teach the applications of science came in 1847. At that time, the university established a school of arts, and appointed James C. Booth "Professor of Chemistry as applied to the Arts," a subject he had previously taught at the Franklin Institute and Central High School.[35] His appointment was a feeble beginning; no other professors were appointed, and Booth had to depend on student fees for his compensation. But more serious plans were under way, primarily due to the efforts of John F. Frazer, professor of natural philosophy and chemistry. Frazer had studied at the University of Pennsylvania under Alexander Dallas Bache and had taught at Central High while Bache was principal there. When Bache moved to the Coast Survey, Frazer took his chair at the University. Frazer was also editor of the *Journal of the Franklin Institute* for many years, became provost of the University, and was later one of the incorporators of the National Academy of Sciences. He was, then, prominent in the inner councils of American science, as well as in local circles.[36]

Early in 1852, Frazer wrote to Charles E. Smith, asking for his ideas on technical education. Smith was president of the Philadelphia and Reading Railroad and was actively interested in the state's iron industry. More to the point, he had traveled extensively in Great Britain and Europe, visiting industrial plants and technical schools. Even before his tour, he had been impressed with the need for a special school for mining and metallurgical engineers, but he returned absolutely convinced of the idea, and devised a plan for a school of mines to be attached to "one of our public institutions." Smith was also certain that Philadelphia, as the center of America's coal and iron mining, its machinery industries, and the financial hub of those interests, was the logical place for such a school.[37] Since their concerns were obviously compatible, Smith outlined his plan to Frazer. To urge its consideration, he appended a brief in support of

---

35. Sack, *Higher Education in Pennsylvania*, II, 465.
36. For information on Frazer, see *Journal of the Franklin Institute*, XCV (1873), 211.
37. Charles E. Smith to Frazer, March 8, 1852, University of Pennsylvania Archives, Philadelphia, Pa.

a technical school, which was signed by twenty-seven of the state's leading iron manufacturers and industrialists.[38]

Smith's plan became the basis for a reorganized School of Mines, Arts and Manufactures. In the outline which Frazer submitted to the chairman of the University's committee on the matter, he added to Smith's course of studies a department of "Theoretical and Practical Mechanics and its application to Machinery." Otherwise, the outline was identical to Smith's and became the program for the new school. The possibility of aid from industry was a persuasive stimulus to the establishment of a school and the formation of its curriculum.[39] Moreover, industry's interests were shared by the faculty. Fairman Rogers, professor of civil engineering, spoke to that point when he described the school's value to Pennsylvania:

> While such inexhaustible beds of coal and iron lay yet undisturbed beneath her surface there is great need of a body of able men to turn her resources to the best advantage and to conduct the factories which are ever increasing within her limits.[40]

Clearly, the school had the opportunity to fill that need.

The School of Mines, Arts and Manufactures also served other needs, as the University sought a firmer foundation for its own interests. Like many of the country's universities, Pennsylvania was faced with a mystifying decline in enrollment in the 1850s. A committee was formed to look into the matter and to make recommendations. Some, expressing a current view, suggested that undergraduate courses be made elective, and thus more attractive. Bishop Alonzo Potter, one of the University's trustees, thundered at such weak-kneed solutions, which would make education "more flimsy and superficial than it now is." What the age demanded was a "real university for men," of the kind proposed at Albany in an 1851

---

38. *Ibid.* The signers included Alan Wood, of the Delaware Iron Works, M. W. Baldwin, of locomotive fame, and Frederick Fraley, president of the Schuylkill Navigation Company and a central figure in Philadelphia's scientific societies. Seventeen of the twenty-seven were iron manufacturers.

39. John F. Frazer to Joseph R. Ingersoll, April 13, 1852, University of Pennsylvania Archives.

40. Fairman Rogers, "Historical Sketch of the School of Mines, University of Pa.," University of Pennsylvania Archives.

meeting of the American Association for the Advancement of Science by Bache, Agassiz, Dana, Peirce, and Frazer. Potter argued for vigorous training and for prizes, scholarships, and fellowships to secure "the highest attainable excellence." [41]

Alexander Dallas Bache, writing from a Coast Survey camp, seconded the bishop's argument: "This, I agree with you, is the time for such a movement, and Philadelphia the best place for it." For those interested in the pure sciences, Bache suggested an even more dazzling prospect:

> Philadelphia is so desirable as a residence, and affords such advantages in a scientific and literary point of view, that if permanent and adequate salaries could be offered, you would be able to gather the elite of the country into the University, and make it the centre of literary and scientific attraction for the whole country. [42]

Money, of course, was the problem, and Bache asked, "Will not some rich Philadelphian give you the endowment necessary?"

Bache's proposal was a vision of grand scale, but no patron stepped forward. In any event, just like the Albany plan, the dream reflected the aspirations of scientists, not the needs of technical education. Without a firm university commitment to the subject, the professors in the School of Mines, Arts and Manufactures continued to give their courses outside of regular teaching duties and still depended on student fees for compensation. This situation, which had been viewed as only a temporary expedient at the school's founding, existed for another decade. Technical education in the University was completely subordinated to the arts and sciences department. Fewer courses were required for the Bachelor of Science degree, the tuition was less, and the classes were unconnected with ordinary classroom instruction.

Only after the Civil War did the requirements of the technical curriculum achieve status equal to those of the Bachelor of Arts

---

41. Alonzo Potter to Joseph R. Ingersoll, Philadelphia, July 8, 1852, *Documents U. of P.*, pp. 3–4, University of Pennsylvania Archives. See A. Hunter Dupree, *Science in the Federal Government* (Cambridge, Mass., 1957), pp. 115–19.

42. A. Dallas Bache to A. Potter, August 2, 1852, *Documents U. of P.*, p. 32.

degree. Disappointed in its hopes to receive the land-grant funds, the University's ability to convey technical education on any larger scale waited for private benevolence. When adequate funding finally came in 1872, Bache was gone, and so was any idea of a national university. Despite its established position and apparently ready access to financial aid, the University of Pennsylvania's efforts at technical education were marked by slow growth and a struggle almost as painful as the one Kennedy's polytechnic endured. And in 1872, even as the University inaugurated a new building in West Philadelphia, the whole point of formal education in the practical sciences was challenged.

A SEPARATE BUT EQUAL EDUCATION

One of the major reasons for relocating the campus had been to revivify the School of Mines, Arts and Manufactures, and the roster of speakers at the inauguration included some of the city's most prominent industrialists. In the most surprising address on that day of celebration, Joseph Harrison, the locomotive manufacturer, told his audience that the effort had been wasted. The University of Pennsylvania was a fine place, he maintained, for the idle rich and for the man who would find pleasure, as well as a profession, in abstract knowledge. But for the great mass of society, those who had to work for their daily bread, "it is the education of the workshop, and not the education of the schools, that is more required." The best knowledge, according to Harrison, was that which could be turned "to the best and most profitable account." It was not theoretical, nor could it be found in the high schools, universities, or even polytechnics with their laboratories and workshops. Native talents, hard work, and practical experience were the only requirements for success.[43]

Coleman Sellers, a distinguished mechanical engineer and president of the Franklin Institute, also considered the question of technical education. Addressing himself directly to Harrison's argument,

---

43. *Proceedings at the Public Inauguration of the Building erected for the Departments of Arts and Science, October 11, 1872* (Philadelphia, Pa., 1872), p. 62.

Sellers claimed that there were two great virtues in a higher education for engineers. The first was training in scientific principles, especially mathematics. "I have never believed in the 'rule of thumb,'" he said. "I have never believed in that intuitive perception which would enable a person to shape a machine without a knowledge of the laws that govern matter and regulate its durability." Of equal importance to Sellers was learning which went beyond "mere technical knowledge" and which awakened sensitivity to "the poetry of words, of art, and of nature." In short, he argued for an education which would form "cultivated, well-educated men." The world needed both kinds of knowledge, and Sellers concluded, to "great applause," that the university could best provide it.[44]

The idea that the "rough-and-tumble discipline" of the workshop was the better teacher was a popular theme which received ample notice in the technical periodicals of the later nineteenth century.[45] But in the history of technical education, Sellers raised a more interesting point. For those men who emerged as leaders in the field of engineering training, the most important debate was not over the value of theory, but whether or not there should be any connection at all between schools of engineering and schools of liberal arts. In the decades following the Civil War, technical educators became an interest group, with their own values and institutional goals, not the least of which was parity with other disciplines.

On that issue, the reaction of foreign observers is revealing. In 1871, the province of Ontario was considering the establishment of a school of technology, and one of the unsettled questions was its relationship to existing educational institutions. For aid in the deliberations, a commission was sent to study technical education in the United States. The commissioners visited Harvard, M.I.T., Worcester Institute, Yale, Columbia, Cornell, Rensselaer, and the Cooper Union. In their report, the Canadians noted:

---

44. *Ibid.*, pp. 80–81.
45. See Monte Calvert, *The Mechanical Engineer in America, 1830–1910* (Baltimore, Md., 1967); Edwin T. Layton, *The Revolt of the Engineers* (Cleveland, Ohio, 1971); and Raymond H. Merritt, *Engineering in American Society, 1850–1875* (Lexington, Ky., 1969).

On no point was the testimony at the Institutions we visited more clear, distinct and uniform than that the proposed School of Practical Science should, in its teaching and management and government, be kept entirely distinct from any other Institution. The more efficient the Institution to which it might be attached, the more certain would be the failure of the School. Even at the two distinguished American Universities of Harvard and Yale, where scientific Schools exist, their efficiency and success is just in proportion to their entire practical separation for teaching and other purposes from the other parts of the University.[46]

For Robert H. Thurston, the passage of time only reinforced his conviction that engineering education should be divorced completely from other types of education. Dean of the Engineering College at Cornell, Thurston was one of the most articulate spokesmen for that view. In an 1893 article outlining the past, present, and future prospects of technical education, he claimed that it was an "almost fatal mistake" to assume that one could get a good general education and at the same time secure proper professional training. Thurston's argument was by analogy to the specialized education required of the other professions. The engineering school should be "a purely professional institution—such as are, for example, all the great law schools—instead of as is customary, attempting to give a mixed course of educational and professional instruction."[47] Thurston's description of the aims of thoroughly specialized instruction is itself instructive. Technical education should advance the student "up into the departments of research or real professional work in applied science."[48] It should provide for the advancement of knowledge as well as its diffusion.

In American technology, just as in American science, the advancement of knowledge meant rigid professionalism, specialization, and research. To achieve those ends, technical educators proposed institutions much like those for which science's elite had yearned.

---

46. J. George Hodgins, *Historical and Other Papers and Documents Illustrative of the Educational System of Ontario, 1856–1872* (Toronto, 1911), II, 189.

47. Robert H. Thurston, "Technical Education in the United States," *Transactions of the American Society of Mechanical Engineers*, XIV (1893), 942.

48. *Ibid.*, p. 943.

Thurston's ideal was a completely integrated system of technological education. At the base was an elementary school, using "reading, writing and spelling books in which the terms peculiar to the trades, the methods of operation, and the technics, of the industrial arts should be given prominence, to the exclusion if necessary of words, phrases, and reading matter of less essential importance to them." [49] Intermediate schools, in this system, would provide more advanced vocational training.

The capstone of his plan was a polytechnic university which would provide "thoroughly scientific training and education," on a level that would place its students "in the front rank among those who do the great work of the profession." But there was more to the university than teaching. Thurston's polytechnic would supply the nation with able teachers, trained investigators, and talented administrators. It would conduct industrial research and coordinate the nation's industrial advance. And perhaps its highest aim would be a technological brain center, "capable of instructing not only the youth . . . but the legislators and executive officers of the government." [50]

Similar ideas were advanced by those who called for a national academy of engineering. All such plans proposed agencies for advanced technical education and research, high standards of performance, government support, and federal policy-making. [51] Earlier in the century, when scientists looked in the same directions, one of their motivations had been to raise American science to the European level. Technical educators felt the same impulse but were also moved by imputations of inferiority at home. Their courses were frequently subordinated to the arts and sciences, and their departments often subverted by the ambitions of scientists. In that respect, the University of Pennsylvania was little different than the Lawrence Scientific School, which Louis Agassiz diverted more to his own ends

---

49. *Ibid.*, p. 857.
50. *Ibid.*, p. 964.
51. William Kent, "Proposal for an Academy of Engineering," *Van Nostrand's Engineering Magazine*, XXXV (1886), 278. For a later example of the same idea, see John Waddell's proposal for the American Academy of Engineers, published in Frank W. Skinner, *Memoirs and Addresses of Two Decades by Dr. J. A. L. Waddell* (Easton, Pa., 1928), p. 103.

than the wishes of its patron. Both William B. Rogers and Francis A. Walker, in their long struggles to keep M.I.T. free of Harvard's deadly embrace, were well aware of university prejudices against technical education. Independence was clearly preferable to amalgamation by "schools where snobbishness makes odious comparisons."[52] As a function of their own needs for status, engineering educators dreamed of prestigious but separate technical schools.

Robert Thurston felt the need for separate engineering schools, but as dean of a land-grant institution, his response was somewhat ambivalent. The Morrill Act represented to him "one of the grandest examples of statesmanlike legislation the world has yet seen," and he could rejoice in the general expansion of technical education.[53] Yet, at the same time, he shared the view that the attempt to blend the classics and the technics in those schools had not "furthered the interests of education in the way that legislators had anticipated."[54] Thurston could really feel comfortable only about those institutions in which strong, independent technical departments pursued purely professional training.

Even within a context of highly specialized training, as distinct from that other kind of education which aimed to socialize man, technical educators still invoked the ideology of democracy. In a fashion reminiscent of an earlier day, they still saw technology as the mechanism for a democratic society, as a broad avenue to a bright future. Wealth from industrial progress was the key. "It is only by the accumulation of wealth," Thurston argued, "that the world can secure the blessings of intellectual or even moral advancement, the comforts of life and healthful luxuries."[55] But material progress depended on education; it therefore became the duty of the state to provide the most effective instruction possible to all of its people. "When manual training and trade schools are found in every town, technical schools in every city, and colleges of science and art in every state," then, Thurston claimed, and only then, would the citizen enjoy "all that he may rightfully demand in his pursuit of every-

---

52. James P. Munroe, *A Life of Francis A. Walker* (New York, 1923), p. 231.
53. Thurston, "Technical Education," p. 915.
54. *Ibid.*, p. 861.
55. *Ibid.*, p. 862.

thing that life and liberty can offer him, and the most perfect happiness that can come to man."[56] Thurston's proclamation was a new declaration of independence, writ modern for industrial America.

The formula was as familiar in 1894 as in 1824. But if the rhetoric had not changed, reality had. At the beginning of the century, technical education was vitally related to the major social, political, and economic questions of the day. Men of such disparate politics as Thomas Jefferson and Joel Barlow could both argue that science—in the broadest sense—education, and democracy "would unite to create a nation in which peace and progress would rule."[57] But by the century's end, technical training was singlemindedly directed toward the production of wealth, in a world increasingly divided by economic competition. Out of the demands of industry and their own systems of values, technical educators constructed a rigorous course of university-level studies, which were separated from most other educational influences and limited in objective. Instead of a broad vista of opportunity within an egalitarian context, technical education contracted into highly specialized training directed toward socially stratified occupational lines. The expertise of management-oriented engineers was shaped into an extremely potent instrument for the creation of industrial wealth. Older ideological dreams, however, were replaced by visions of increased production. Useful knowledge once had a democratic ring; by the end of the century practical meant profitable. As a result, technical education in America lost the power of its earlier message and some of its ability to deal with dynamic social and political issues.

---

56. *Ibid.*, p. 863.
57. Arthur A. Ekirch, *The Idea of Progress in America, 1815–1860* (New York, 1951), pp. 29–32.

Contributors

MARK BEACH
is assistant professor of history and education and associate dean of the
College of Arts and Science at the University of Rochester.

ROBERT V. BRUCE
is professor of history at Boston University.

GEORGE H. DANIELS
is professor of history at Northwestern University.

ROBERT C. DAVIS
is associate professor of sociology at Case Western Reserve University.

A. HUNTER DUPREE
is George L. Littlefield Professor of History at Brown University.

DANIEL J. KEVLES
is associate professor of history at the California Institute of Technology.

EDWIN LAYTON
is associate professor of the history of science and technology at Case
Western Reserve University.

EDWARD LURIE
is professor of health sciences and history at the University of Delaware.

HOWARD S. MILLER
is associate professor of history at the University of Missouri, St. Louis.

CARROLL PURSELL

is associate professor of history at the University of California, Santa Barbara.

NATHAN REINGOLD

is editor of the Joseph Henry Papers at the Smithsonian Institution.

CHARLES E. ROSENBERG

is professor of history at the University of Pennsylvania.

BRUCE SINCLAIR

is associate professor at the Institute for the History and Philosophy of Science and Technology at the University of Toronto.